Schools Council
Educational Use of Living Organisms
General Editor: P J Kelly

Plants

Author: C D Bingham
Contributor: J D Wray

HODDER AND STOUGHTON

Schools Council Educational Use of Living Organisms Project
Director: P J Kelly
Research Fellow: J D Wray

This project was established at the Centre for Science Education, Chelsea College, in 1969. Its main aims have been to determine the needs of schools with respect to living organisms, to evaluate the usefulness of various kinds of organisms for educational purposes, and to devise maintenance techniques and teaching procedures for the effective use of appropriate species.

ISBN 0 340 17053 0

First published 1977

Hodder and Stoughton Educational,
a division of Hodder and Stoughton Limited,
Mill Road, Dunton Green, Sevenoaks,
Kent

Printed and bound in Great Britain by
Unwin Brothers Limited

Computer Typesetting by Print Origination, Bootle,
Merseyside, L20 6NS

Preface

The use of living organisms in schools, while not new, has received considerable emphasis in recent developments in the teaching of biology, environmental studies and allied subjects in colleges and secondary schools, and in many aspects of primary school work. It is a change reflecting the much wider movement in educational thinking which acknowledges both the interest and delight that young people can gain from animals and plants and the importance of fostering an appreciation of the scientific, social, aesthetic and moral issues involved in the study of life and the natural environment. It is a change, also, that presents some very real – but not insurmountable – practical problems for schools.

The implications of using living organisms for education are basically threefold. There needs to be an adequate supply of appropriate organisms and they should be kept healthy and be able to live as naturally as possible. Accommodation for the organisms should allow them to be observed and studied easily but with respect. Maintenance of the organisms should not place an undue burden on teachers and technicians. The Educational Use of Living Organisms publications are intended to assist teachers (including, of course, student teachers), technicians and administrators to contend with these implications.

The books in the series deal with the principles which underlie the effective educational use of living things. They also provide information to help teachers integrate work with organisms into their courses and to cope with the practical, day-to-day problems involved. For teachers and administrators there are technical details of value for planning facilities, and annotated bibliographies provide the guidelines for more detailed studies if they are required. In addition, posters and slide transparencies for use with pupils have been produced complementary to the books.

The series has been produced as part of the work of the Educational Use of Living Organisms project. The project was initiated by the Institute of Biology and Royal Society Biological Education Committee and was established as a Schools Council Project at the Centre for Science Education, Chelsea College (University of London), in 1969. While the project's major financial support was from the Schools Council, the Nuffield Foundation and Harris Biological Supplies have also given generous contributions.

Many people have assisted in a personal capacity and we have particular regard to the interest shown by Mr J A Barker, Mr D J B Copp, Mr T A Gerrard, Mr J F Haller, Mr O J E Pullen and Dr C H Selby.

Mr J D Wray undertook much of the research of the project and provided much of the information on which the books are based. In this, he was admirably supported by Mr J B Green as technician and Miss M Hoy and later Mrs E Barton as the project's secretary.

The authors of the books very kindly have been most tolerant of their editor. I would like to express my gratitude to them and our publisher for both their help and forbearance.

P J Kelly
Director, Educational Use of Living Organisms Project

Contents

Introduction

It is unusual to find a school in which there are no growing plants. They are to be found in school grounds, in classrooms and laboratories, even in staffrooms. They are decorative and can be used in a wide variety of activities from art to the study of living things. Moreover, most are easy to grow.

Suitable pot plants are amongst the easiest of all living things for children to keep and to be responsible for maintaining (Fig. 1). Children find it enjoyable to grow these plants, especially if they will flower. They can watch the development from seed or cutting to maturity, observing a complete life-cycle and many fundamental living processes. During holiday periods these plants can be taken home to be cared for with little domestic upset.

Keeping plants has its own difficulties and problems and those suitable for schools are no exception. It is the purpose of this book to show how to grow plants well, how to minimise any problems which can arise and how to use plants effectively.

Clearly the choice of plants and facilities is critical. A list of the most suitable plants is given on pages 8 to 18, indicating where they are best grown and how they are propagated and used. This list is not exhaustive, many other plant species or cultivated varieties (cultivars) are, or can be, kept in schools. Those named are compact, and easy to maintain by anyone, even beginners. They are relatively cheap and can be obtained from the suppliers given in Appendix 2, many may also be available locally.

Some of these plants can be kept perfectly well in classrooms or laboratories (Chapter 1); others are best grown out of doors (Chapter 2) or in glasshouses (Chapter 3). These facilities are letter coded as indicated. A key is given on page 8, thus those with a letter A can be grown in the classroom with no additional facilities and so on.

The necessary cultural techniques for maintenance and propagation of the listed liverworts and mosses (Bryophytes), ferns (Pteridophytes), conifers (Gymnosperms) and flowering plants (Angiosperms) are described in Chapter 4. The culture and use of algae, fungi and bacteria (fungi and bacteria are not classified as plants by some authorities) is dealt with in another of the books in the Educational Use of Living Organisms series, 'Micro-organisms' and of plants for genetical use in 'Organisms for Genetics'.[99]

The variety of educational uses is described in Chapter 5 in which it is possible to find the most suitable plant for a particular use. The main uses for each plant are also given in the plant list enabling their potential value to be readily assessed.

Those schools having access to the natural vegetation of the countryside, parkland or garden facilities can utilise plants grown in the normal climatic conditions out of doors to provide materials for much of the work requiring plant material in the laboratory or elsewhere in the school. For example, twig structure, leaf shapes and development, flower structure, inflorescence types, fruits of varying structure and mode of dispersal, and root systems can all be covered using natural material or crops and weeds from gardens. Such material can also be used in the art and craft department and for decoration. The major drawback is due to the seasonal rhythm, since it happens that for us the most interesting inflorescences, flower and fruit structures inevitably appear late in the summer term or during the long summer holiday. It is partly for this reason, and also to increase the range of subjects, that some form of protected cultivation must be introduced.

When suitable material is collected from natural habitats, the conservation of each species must be carefully considered.[14] A most important aspect in contemporary education is that pupils should be made aware of the need for the conservation of many species by avoiding the removal of excessive numbers of plants, flowers or fruits. The destruction or serious change of habitats, for example by heavy trampling, can also seriously reduce population numbers. Indiscriminate uprooting of wild plants is to be strongly discouraged and plants known to be rare, for example all the British orchids, should not be collected.[2,11]

Finally for those not familiar with technical terms, which have been deliberately kept to a minimum, there is a glossary giving an explanation on pages 5 to 7.

Glossary

Adventitious roots—roots which develop in an abnormal place e.g. at the base of a stem taken as a cutting (qv).

Annual—a plant which completes its life cycle involving seed germination, flowering, seed production and death of the mature plant in one year e.g. *Calendula.*

'Bark-bound'—woody (qv) plants which have grown slowly or suffered a check resulting in the bark hardening and restricting further growth. Continued growth is possible only if the bark is broken or cut.

Biennial—a plant which requires two years to complete its life cycle e.g. carrot.

Bulb—a storage and perennation (qv) organ formed from swollen leaf scales e.g. *Narcissus,* Hyacinth.

Calcicole—a plant which does not thrive in acid soils preferring alkaline soils e.g. *Dianthus.*

Calcifuge—a plant which does not thrive in soils containing much chalk or lime or when watered with alkaline water, preferring acid soils e.g. heaths, *Rhododendron.*

Carpel—the part of a flower consisting of ovary, style and stigma (qv).

Cladode—a flattened structure formed from a stem but looking like and functioning similarly to a leaf e.g. the flattened parts of many cacti.

Composite inflorescence (qv) or 'flower'—a 'head' with many flowers closely packed on a button shaped terminal part of the stem. This is characteristic of plants of the daisy family.

Compost—either a material for growing plants in consisting of a special mixture of loam, peat, sand, and appropriate plant nutrients (see soil-less medium) or rotting vegetation used for incorporating in soil or for top dressing.

Corm—a storage and perennation (qv) organ formed from the swollen base of a stem e.g. *Crocus, Gladiolus.*

Cutting—a detached portion of a plant used for propagation from which roots develop.

Cyme—an inflorescence (qv) in which the tip or terminal bud flowers first. Further growth occurs from lower down the stem which in turn produces terminal flowers e.g. *Ranunculus.*

'Double digging'—a method of cultivating the soil in which a trench is opened up with a spade and ultimately replaced. The lower soil in each trench is loosely forked to improve the drainage and compost (qv) or manure usually incorporated. This is covered by the material from the next trench which is turned onto it.

Dicotyledon—flowering plants with two seed leaves, flower parts usually in 4's or 5's and leaf veins branching from a central vein or mid-rib e.g. roses.

Dioecious—unisexual plants bearing male flowers separate from those bearing female flowers e.g. Holly.

Disease—a condition of ill-health suffered by living things. In plants the cause may be due to a variety of factors including the presence of certain chemicals in their environment, the absence of certain nutrient elements, inadequate water supply, excessive transpiration (qv) or the activities of other living things.

Endospermic—a seed (qv) in which food is stored within the endosperm e.g. maize, castor oil.

Ephemeral—a plant which completes its life cycle in a very short period of time e.g. Shepherd's purse.

Epigeal—the germination of a seed during which the seed leaves or cotyledons are forced above ground e.g. radish, runner beans.

Fertilisation—the fusion of reproductive cells. In flowering plants this takes place within the ovule (qv) resulting in the formation of an embryo inside a developing seed (qv).

Fruit—the structure formed from the ovary (qv) of a flowering plant after fertilisation and which contains the seeds.

Grafting—the joining of one part of a plant (the scion) to a rooted stock by the junction of growing tissue. Grafts can join stem to stem, stem to root, or bud to stem (budding).

'Hardening off'—the process of gradually accustoming plants to lower temperatures and humidity by moving them from the glasshouse to unheated frames and then progressively increasing the ventilation.

Herbaceous—a plant with a stem which remains soft and green and which dies back to ground level each year e.g. *Calendula*.

Hormones—a group of plant growth promoting or inhibiting substances. Some are produced naturally e.g. auxins; others are synthetic chemicals e.g. gibberellic acid. They are used for example as an aid to the development of roots in cuttings (qv) or as weed killers.

Hypogeal—the germination of a seed during which the seed-leaves or cotyledons remain below ground e.g. broad bean, castor oil.

Inflorescence—the part of a plant bearing more than one flower. *See* composite, cyme, raceme and umbel.

Loam—a type of soil good for the growth of most plants consisting of a balanced mixture of mineral particles with plenty of humus. Steam sterilised loams are the basis of the John Innes Composts.

Monocarpic—a plant which flowers once in its life cycle, sets seeds and then dies. Most true annual (qv) or ephemeral (qv) plants including shepherd's purse, groundsel and some long-lived plants such as bamboo.

Monocotyledon—flowering plants with a single seed leaf, flower parts usually in 3's and parallel leaf veins e.g. *Iris*.

Monoecious—plants in which male and female flowers are on one plant. The male and female parts may be in one flower (hermaphrodite) or in separate flowers on the same plant.

Offset—a short runner (qv). Roots and a new plant develop at the tip.

Ovary—that portion of a flower either separate or fused with others, in which the ovules (qv) and then the developing seeds are protected.

Ovule—small structures which, after fertilisation, develop into seeds. In flowering plants they are contained within the ovary.

Perennation—the survival of a plant from season to season usually involving an intervening period of little activity or dormancy.

Perennial—a polycarpic (qv) plant which flowers year after year e.g. *Narcissus*, or a monocarpic (qv) plant which takes a long time to mature before flowering e.g. bamboo.

Pest—a living thing which attacks plants e.g. mildew fungus and greenfly.

pH—a term used to express the acidity or alkalinity of a solution in water. The scale is from 1 to 14, 1 being the most acidic, 14 being the most alkaline, 7 being neutral. pH may be measured electrically or by the use of coloured indicators.

Phylloclade—a flattened structure formed from a leaf stalk (petiole) looking like and functioning similarly to a leaf e.g. *Acacia*.

Pollination—the transference of pollen grains from the stamens (qv) to the stigma (qv) of the same flower or a flower on the same plant (self-pollination) or of a flower on another plant of the same species (cross-pollination).

'Pot-bound'—plants growing in pots in which their roots are so tightly packed that it is difficult to get water into the soil/root ball. Such conditions stunt growth but often force flowering.

'Potting on'—transplanting growing plants from a seed tray to a plant pot or from a small plant pot to a larger one.

'Pricking out'—the careful separation of seedlings which are growing closely together and planting them, usually in rows, in boxes or seed trays about 4-8 cm apart.

Protandrous—flowers where the stamens (qv) ripen and shed pollen grains before the stigma (qv) is receptive e.g. *Antirrhinum.*

Protogynous—flowers where the stigma (qv) ripens before the stamens (qv) shed pollen grains e.g. radish.

Raceme—an inflorescence (qv) which grows continuously from the tip. Flowers are continuously formed, the lowest flowering first e.g. *Antirrhinum.*

Runner—a prostrate shoot; at the tip, roots and a new plant develop e.g. strawberry.

Seed—the structure formed from a fertilised ovule (qv) containing an embryo which can grow into a new plant on germination.

'Soil-less' medium or compost—a material for growing plants in, consisting of a mixture of peat, sand and appropriate plant nutrients.

Stamens—the parts of flowers consisting of a stalk and anther in which the pollen grains develop.

Stigma—the part of the ovary (qv) of a flower to which pollen grains adhere.

Stolon—a stem with buds and scale leaves. From the tip, roots and a new plant develop when it touches the ground.

Sucker—a shoot which arises below ground. The tip appears above ground some distance from the main plant, e.g. mint.

Systemic pesticides—pesticides which enter and remain for a long time in the sap of the plant destroying internal pests and insects which obtain the sap e.g. greenflies or which eat the plant e.g. caterpillars. These substances may be toxic to people and domestic animals.

Transpiration—the loss of water vapour from the aerial parts of a plant, especially the leaves. This loss is normally balanced by the intake of water by the roots.

Transplanting—moving plants from one situation to another. It is best done when they are dormant or when the plants are young seedlings or when new roots are being produced often just after flowering e.g. primulas and many bulbs.

Tuber—a storage and perennation (qv) organ formed from a swollen root e.g. *Dahlia,* celandine, or stem e.g. *Iris,* potato.

Umbel—an inflorescence (qv) with all the flower-stalks arising from one point of the stem. This is characteristic of plants of the carrot family.

Woody—a plant with a stem in which woody tissue develops, and which remains above ground e.g. *Rosa* and trees.

List of Suitable Plants

This list gives details of the plants most suitable for use in schools. Any of these may be grown with confidence, providing the necessary facilities are available and the cultural directions are followed.

Plants are listed, in alphabetical order, by generic name. This name is always used whenever a plant is mentioned in the text. Each entry gives the name of a particular species, or the abbreviation spp. is added to the generic name to indicate that a number of species or their cultivated varieties (cultivars) are suitable. The generic or species name is followed by the common name, unless it is the same and then the family to which the plant belongs.

The facilities necessary to maintain the plant are indicated by a letter code which is given below. These facilities are discussed in Chapters 1, 2 and 3.

Chapter 1, pages 19 to 25.

A Classroom or laboratory with no additional facilities.

B Laboratory with some facilities.

Chapter 2, pages 26 to 29.

C Outdoors, unprotected in the open—the garden. This is subdivided depending on the soil type, its acidity or alkalinity and the degree of moisture.

	Acid $(pH < 7)$	*Neutral* $(pH = 7)$	*Alkaline or calcareous* $(pH > 7)$
Sandy	Csa	Csn	Csc
Loam	Cla	Cln	Clc
Clay	Cca	Ccn	Ccc
Marshy	Cma	Cmn	Cmc
Pond	Cpa	Cpn	Cpc

D Outdoors, protected by glass or plastic covers, which may be movable, or by a bird-proof frame.

Chapter 3, pages 30 to 36

E The fixed glasshouse or greenhouse.
This is subdivided according to the heating available.

Unheated	Eu
Heated to be 'frost-free' with a minimum temperature of 0°C (32°F)	Ef
Heated to keep a minimum temperature of 7°C (45°F) in the winter	Eh

The methods of propagation are given next. The least easy method(s) are bracketed. Full details of the procedures are given in Chapter 4, pages 37 to 48.

Finally the main uses of each plant are indicated by a number code which is given below. These uses are discussed in Chapter 5, pages 49 to 63.

Biological enquiry.

1 *Physiology*
- .1 Germination
- .2 Growth
- .3 Basic properties of protoplasm
- .4 Sensitivity and Tropisms
- .5 Transpiration
- .6 Nutritional requirements
- .7 Photosynthesis
- .8 Respiration
- .9 Enzyme action
- .10 Food storage

2 *Morphology*
- .1 Bryophyte
- .2 Pteridophyta and Gymnospermae
- .3 Angiospermae, Monocotyledons and Dicotyledons
- .4 Structure of root stem and leaf
- .5 Special adaptations
- .6 Flower structure
- .7 Dispersal—seed and fruit; spore

3 *Anatomy and microscopic structure*
- .1 Root
- .2 Stem
- .3 Special conducting tissues
- .4 Leaf
- .5 Spore production and reproductive bodies
- .6 Flower structure
- .7 Squashes to show mitosis or meiosis

4 *Life cycles and perennation*
.1 Annual, biennial, perennial, monocarpic
.2 Pollination

5 *Propagation*
.1 Vegetative
.2 Artificial

6 *Systematic collections*
.1 Order beds
.2 Variation
.3 Special hybrids
.4 Diploid triploid and polyploid plants
.5 An extensive list of plants for genetics will be found in the companion volume 'Organisms for Genetics'.[99]

Centres of Interest and Activity

7 *'Arts'*
.1 Leaf prints
.2 Plaster casts
.3 Plants to dry as specimens or for decoration
.4 Plants with decorative foliage

8 *'Crafts'*
.1 Coppice
.2 Osiers and willows
.3 Timber

Associated Work in other Subjects

9 *Economic plants*
.1 Market garden crops
.2 Farm plants
.3 Herbs
.4 Medicinal plants
.5 Tropical plants

Remedial and other Beneficial Uses

10 Suitable plants

Abies spp. (see also *Picea*). Silver fir.
Gymnospermae, Pinaceae.
Species include some dwarf forms suitable for growing in pots or on rockeries e.g. *A. albertiana conica.*[31]
Facilities Csa, Cla or Cln
Propagation Seed
Use 2.2; 3.2; 3.3; 3.5

Abutilon hybridum and *A. vitifolium.* Malvaceae.
Variegated form has clear distribution between areas with and without chlorophyll.
Facilities A, B, E (Cln in summer only)
Propagation Seed/stem cuttings
Use 1.7; 7.4

Acacia spp. Mimosa. Leguminosae.
Transition from phylloclades to pinnate leaves.
Facilities B, Eh
Propagation Seed (stem cuttings)
Use 2.5

Acer spp. Sycamore, maple. Aceraceae.
Some Japanese maples have interesting leaf shapes and pigments.
Facilities Cln
Propagation Seed (grafting/budding/layering)
Use 1.1; 7.1; 7.4; 8.3

Achillea spp. Milfoil, yarrow. Compositae.
Many interesting species with aromatic qualities useful as herbs. Interesting composite inflorescences.
Facilities Cln
Propagation Seed/division
Use 2.6; 5.1; 9.3

Aconitum spp. Monkshood. Ranunculaceae.
Facilities Cln
Propagation Seed/division
Use 2.6; 6.1; 9.4

Acorus calamus. Sweet flag. Araceae.
Facilities Cmn, Cpn
Propagation Division
Use 2.4

Aesculus hippocastanum. Horse-chestnut.
Hippocastanaceae.
A. carnea, the red flowered variety, is a tetraploid. Plants grown from seeds (conkers) kept indoors in plant pots provide useful bud opening sequences in the winter.
Facilities Cln
Propagation Seed/layering/grafting
Use 1.1; 2.4; 8.3

Agave americana. American aloe and other *Agave* spp. Amaryllidaceae.
Good example of a monocarpic plant. Some species take up to 50 years to flower.

Facilities B, Eh
Propagation Offsets
Use 2.5; 4.1

Alisma plantago-aquatica. Water plantain. Alismaceae.
Facilities Cpn, Cpc
Propagation Division (seed)
Use 2.5; 3.3

Allium spp. Onion, leek, chives. Liliaceae.
Many species used as vegetables or herbs. Some ornamental flowering species used in rockeries.
Facilities Cln
Propagation Seed or division
Use 1.1; 1.3; 1.4; 2.5; 2.6; 3.1; 3.4; 3.7; 4.1; 5.1; 9.1; 9.3

Aloe spp. Liliaceae.
Many ornamental species showing succulent and xerophytic modification.
Facilities A, B, Eh
Propagation Seed
Use 2.5

Ampelopsis vitis (syn. *Parthenocissus*). Virginian creeper. Vitaceae.
Useful to show modification of leaves to form suction pads for climbing.
Facilities Cln
Propagation Cuttings
Use 2.5

Anaphalis spp. Everlasting flower. Compositae.
Facilities Cln
Propagation Seed
Use 7.3

Antirrhinum spp. Snapdragon. Scrophulariaceae.
Facilities Cln
Propagation Seed/cuttings
Use 2.6; 2.7; 6.3; 6.4

Apium graveolens. Celery. Umbelliferae.
Food storage in petiole.
Facilities Cln
Propagation Seed
Use 3.4; 9.1

Aquilegia spp. Columbine. Ranunculaceae.
Facilities Cln
Propagation Seed/division
Use 2.6; 2.7; 6.1

Arum spp. Araceae.
Facilities Cln, Ef
Propagation Seed/division
Use 4.2

Arundinaria spp. Bamboo. Graminae.
Monocarpic.
Facilities Cln, Cmn
Propagation Division
Use 2.5; 3.2; 4.1; 7.3; 8.2 (split cane)

Arundo donax. Reed. Graminae.
Special pith cells.
Facilities Cmn, Cpn
Propagation Division
Use 3.2; 3.3; 8.2

Avena spp. Oat. Graminae.
Facilities Cln
Propagation Seed
Use 1.2; 1.6; 2.6; 2.7; 7.3; 9.2

Begonia spp.[73] Begoniaceae.
Facilities A, B, Ef, (Cln in summer)
Propagation 'Rex' begonias by leaf cuttings, tuberous species by division, others by seed
Use 7.4; 10

Beta vulgaris. Beet-root. Chenopodiaceae.
Facilities Cln, Csn
Propagation Seed
Use 1.3; 1.10; 2.5; 4.1; 6.4; 9.1

Bletia (Syn *Bletilla*) spp. Ground orchids. Orchidaceae.
Facilities B, Ef
Propagation Division
Use 2.6

Brassica spp. Broccoli, cabbage, cauliflower, mustard, turnip, sprout. Cruciferae.
Facilities Cln, Clc
Propagation Seed
Use 1.1; 2.4; 2.5; 3.1; 6.2; 6.3; 9.1; 9.2

Bryophyllum daigremontianum and *B. tubifolium* (syn. *Kalanchoe*).[73]
New plantlets produced along the leaf edges.
Facilities A, B, Eh
Propagation Seed/plantlets
Use 2.5; 5.1

Buxus spp. Box. Buxaceae.
Facilities Cln
Propagation Stem cuttings/division
Use 3.4

Calendula officinalis.[69] Pot marigold. Compositae.
Facilities A, B, Eh as pot plants or Cln if outdoors
Propagation Seed
Use 2.6; 10

Calluna vulgaris. Ling heather. Ericaceae.
Facilities Csa, Cla, Cca
Propagation Cuttings/division
Use 2.4; 3.4

Cheiranthus cheiri and other species. Wallflower. Cruciferae.
Facilities Cln, Clc
Propagation Seed/stem cuttings
Use 2.3; 2.6

Chlorophytum capense.[73] Spider plant. Liliaceae.
Facilities A, B, Eh
Propagation Offshoots/division (seed)
Use 5.1; 7.4; 10

Chrysanthemum spp.[69] Various hardy species and greenhouse varieties. Compositae.
Facilities Cln, D, Ef
Propagation Perennials by division/stem cuttings/seed. Annuals by seed.
Use 2.6; 6.2

Clematis spp. Virgin's bower. Ranunculaceae.
Facilities Cln, Clc
Propagation Seed/stem cuttings (layering)
Use 2.5; 2.7; 3.3; 6.1; 7.3 (Old man's beard)

Colchicum spp.[87] Autumn crocus. Liliaceae.
Facilities Cln
Propagation Seed/division of bulbs
Use 2.5; 9.4

Coleus blumei & other cultivars. Painted leaf plant, flame nettle. Labiatae.
Facilities A, B, Ef
Propagation Seed/stem cuttings
Use 7.1; 7.4; 10

Conocephalum conicum.[30] Bryophyta, Hepaticae, Marchantiaceae.
Facilities Cln, Eu
Propagation Division/gemmae
Use 1.6; 2.1; 2.7

Cornus spp. Cornell and dogwood. Cornaceae.
Variety of bark colours on the twigs.
Facilities Clc
Propagation Stem cuttings
Use 7.2; 8.2 (coloured wood for basketry)

Corylus spp. Hazel. Betulaceae.
Catkins for wind pollination.
Facilities Cln
Propagation Seed (suckers/layering/grafting)
Use 2.6; 4.2; 7.3; 8.1

Crassula spp.[73] Crassulaceae.
A variety of succulent species.
Facilities A, B, Eh
Propagation Seed/stem cuttings
Use 2.5; 3.4; 5.2

Crataegus spp. Hawthorn. Rosaceae.
Useful intergenic graft hybrid with medlar Crataegomespilus.
Facilities Cln
Propagation Seed, store berries in sand outdoors for a year before sowing
Use 6.3; 7.1; 7.2

Crocus spp.[87] Iridaceae.
Facilities A, B, Eu in pots; Cln
Propagation Seed/division of corm offsets
Use 2.5; 5.1

Cucumis & *Cucurbita* spp. Cucumber, melon. Cucurbitaceae.
Facilities Cln, D, Ef
Propagation Seed
Use 2.4; 3.3; 4.2; 9.1

Dahlia spp. Compositae.
Facilities Cln
Propagation Seed/stem cuttings/division of tubers
Use 1.10; 2.4; 2.5; 4.1; 5.2; 6.2

Datura stramonium. Thorn apple. Solanaceae.
Facilities Csn, Csc, Cln
Propagation Seed
Use 9.4

Daucus carota. Carrot. Umbelliferae.
Facilities Csn, Cln, D
Propagation Seed
Use 2.5; 9.1

Delphinium spp. Larkspur. Ranunculaceae.
Facilities Cln
Propagation Seed/stem cuttings/division
Use 6.1; 7.3

Dianthus spp. Sweet william, carnation, pinks. Caryophyllaceae.
Facilities Cln, Clc, Ef
Propagation Seed/stem cuttings/layering
Use 10

Digitalis purpurea and cultivars. Foxglove. Scrophulariaceae.
Facilities Csn, Csc, Cln
Propagation Seed/division
Use 2.7; 9.4

Dionaea muscipula. Venus fly trap. Droseraceae. Insectivorous.
Facilities B, Eh
Propagation Seed/division
Use 1.4; 1.6; 2.5

Dracaena draco. Dragon tree. Liliaceae.
Monocotyledon with secondary thickening.
Facilities Eh
Propagation Seed/root cuttings
Use 2.3; 3.2

Drosera spp. Sundew. Droseraceae.
Insectivorous.
Facilities B, Cma, Eh
Propagation Seed/division
Use 1.4; 1.6; 2.5

Dryopteris spp. *D. filix-mas.*[64,86] Male fern. Pteridophyta, Polypodiaceae.
Facilities Cs, Cla
Propagation Spores/division
Use 2.2; 2.7; 3.5; 7.4

Elodea canadensis.[62] Canadian pond weed. Hydrocharitaceae.
Facilities A, B, Cpn
Propagation Division
Use 1.3; 1.7; 5.1

Endymion hispanicus (Syn *Scilla*). Bluebell. Liliaceae.
Facilities Cln
Propagation Offsets from bulb (seed)
Use 3.1; 3.2; 3.7

Epilobium spp. (Syn *Chamaenerion*). Willow herb. Onagraceae.
A wide variety of species within the genus.
Facilities Cln
Propagation Seed/division
Use 1.5; 2.7; 4.2

Epiphyllum spp. 'Leaf flowering' cactus. Cactaceae.
Facilities B, Eh
Propagation Stem cuttings/seed
Use 2.5

Equisetum maximum. Horsetail. Pteridophyta, Equisetaceae.
If in pots water freely.
Facilities Cmn
Propagation Division
Use 2.2; 2.7; 3.5

Erica spp. Heaths. Ericaceae. Mycorrhiza.
Facilities Cla, Cca, Eu, Ef
Propagation Stem cuttings/division
Use 2.4; 2.5

Eryngium maritimum. Sea holly. Umbelliferae.
Facilities Csn
Propagation Seed/division
Use 2.5; 7.3

Euphorbia spp.[73] Spurges and succulents. Euphorbiaceae.

Facilities	Cln, Ef, Eh
Propagation	Stem cuttings/division
Use	2.5

E. pulcherrima (syn. *Poinsettia pulcherrima*). Poinsettia, shows quick response to short day length in the colouring of the leaves.

Facilities	A, B, Ef
Propagation	For all species stem cuttings
Use	1.2; 2.5

Fagus sylvatica & varieties. Beech. Fagaceae.

Facilities	Cln
Propagation	Seed
Use	3.2; 7.3; 8.1; 8.3

Fatsia japonica.[73] Fig leaf palm, Large leaf ivy. Araliaceae.
Intergeneric hybrid *Fatshedera* from *Fatsia* x *Hedera.*

Facilities	Cln, Ef
Propagation	Root and stem cuttings
Use	6.3

Ficus spp.[73] Fig, climbing fig. *F.elastica* India rubber plant. Moraceae.
Useful house plants showing a variety of modifications. Good classroom plants.

Facilities	A, B, Eh best
Propagation	Stem cuttings/air layering
Use	5.2; 7.4

Fragaria cultivars. Strawberry. Rosaceae.

Facilities	Cln
Propagation	Separation of runners (seed)
Use	2.5; 2.7; 5.1; 9.1

Fraxinus spp. Ash. Oleaceae.

Facilities	Cln
Propagation	Seed (grafting)
Use	2.4; 2.7; 8.3

Fritillaria spp.[87] Fritillary, Crown imperial, Snakes head. Liliaceae.

Facilities	Cln
Propagation	Seed/offsets
Use	3.7

Fuchsia spp. and cultivars. Onagraceae.

Facilities	A, B, Ef. Hardy varieties Cln
Propagation	Stem cuttings/seed
Use	1.5; 6.2

Funaria hygrometrica.[30] Bryophyta. Musci Funariaceae.

Facilities	B, Cla, Eu
Propagation	Division/spores
Use	2.1; 2.7

Ginkgo biloba. Maidenhair tree. Gymnospermae Ginkgoaceae.

Facilities	Cln
Propagation	Seed
Use	2.2; 3.5 (Dioecious)

Gomphrena globosa. Globe amaranth. Amaranthaceae.

Facilities	Ef
Propagation	Seed
Use	7.3

Gynura aurantiaca. Compositae.

Facilities	A, B, Eh
Propagation	Stem cuttings
Use	7.4

Hedera spp.[73] Ivy. Araliaceae.

Facilities	A, B, Cln
Propagation	Stem cuttings. Adventitious roots along stem.
Use	2.5; 7.2; 7.4

Helianthus tuberosus. Jerusalem artichoke. Other *Helianthus* spp. Sunflower.

Facilities	Cln
Propagation	Seed/division of roots or tubers
Use	1.10; 2.3; 3.2; 5.1; 9.1

Helichrysum spp. Everlasting flower. Compositae.

Facilities	Cln
Propagation	Seed/stem cuttings
Use	7.3

Helipterum spp. (syn. *Acroclinium rhodanthe*). Australian everlasting.

Facilities	Cln, Ef in pots
Propagation	Seed
Use	7.3

Hippuris vulgaris. Mare's tail. Haloragidaceae.
Facilities Cpn
Propagation Division
Use 3.2; 3.3

Hordeum spp. Barley. Graminae.
Some species cultivated as decorative garden plants.
Facilities Cln, D
Propagation Seed
Use 1.9; 2.6; 7.3; 9.2

Humulus lupulus. Hop. Urticaceae.
Facilities Cln, D
Propagation Seed
Use 2.5; 7.4; 9.2

Hyacinthus spp.[87] Hyacinth. Liliaceae.
Facilities A, B, Cln, Ef
Propagation Offsets from bulbs/seed
Use 3.1; 3.2; 5.1; 10

Hydrangea spp. and cultivars.[73] Saxifragaceae.
Facilities Cln, Eu
Propagation Stem cuttings
Use 1.5; 7.3

Ilex spp. Holly. Aquifoliaceae.
Facilities Cln
Propagation Seed/stem cuttings
Use 3.4; 8.3

Impatiens Balsamina. Balsam. Balsaminaceae.
Explosive fruits.
Facilities Cln, Eh
Propagation Seed/stem cuttings
Use 2.7

Iris spp. Iridaceae.
Facilities Cln, Ef
Propagation Division of the rhizomes/seed
Use 1.7; 2.4; 2.5; 3.4; 5.1

Kalanchoe spp. (syn. for some species is *Bryophyllum*). Crassulaceae.
Facilities B, Eh
Propagation Seed/stem cuttings/detaching small plantlets
Use 5.1

Kleinia articulata. Candle plant. Compositae.
Facilities A, B, Eh
Propagation Stem cuttings
Use 2.5

Larix spp. Larch. Gymnospermae, Pinaceae.
Deciduous conifer.
Facilities Cln
Propagation Seed
Use 2.2; 3.2; 3.5; 8.3

Lathyrus odoratus. Sweet pea. Leguminosae.
Facilities Cln, D
Propagation Seed
Use 2.5; 2.6; 2.7; 10

Lemna spp. Duckweed. Lemnaceae.
Facilities Cpn, Eu
Propagation Division
Use 1.2; 1.6; 2.5; 5.1

Lepidium sativum. Cress. Cruciferae.
Facilities A, B, Ef
Propagation Seed
Use 1.4; 2.4

Ligustrum spp. Privet. Oleaceae.
Facilities Cln
Propagation Seed/stem cuttings
Use 1.5; 2.4; 2.5; 3.4

Lilium spp. Lily. Liliaceae.
Facilities Cln, Ef
Propagation Seed/offsets/separation of bulb scales
Use 2.3; 2.5; 2.6; 3.6; 3.7; 5.2; 7.3

Lunaria annua. Honesty. Cruciferae.
Facilities Cln
Propagation Seed
Use 2.7; 7.3

Lycopersicum esculentum. Tomato. Solanaceae.
Facilities Cln, D, Ef
Propagation Seed/grafting
Use 5.2; 6.3; 9.1

Marchantia polymorpha.[30] Bryophyta, Hepaticae, Marchantiaceae.
Facilities Cln, Eu

Propagation Division/spores/asexual reproductive units called gemmae
Use 1.6; 2.1; 2.7

Marsilea spp. Water fern. Pteridophyta, Marsileaceae.
One of the easiest species to grow is *M. Drummondii.* Relatively large motile male gametes are produced.
Facilities Eh, stand shallow pot in water
Propagation Division
Use 3.5

Matthiola spp. Stock. Cruciferae.
Facilities Cln, Ef
Propagation Seed
Use 10

Mentha spp. Mint. Labiatae.
Facilities Cln
Propagation Stem cuttings/separation of runners or suckers
Use 2.5; 5.1; 9.3

Mimosa pudica. Sensitive plant. Leguminosae.
Perennial treated as an annual.
Facilities B, Eh
Propagation Seed
Use 1.4; 2.5

Narcissus spp.[87] Daffodil. Amaryllidaceae.
Facilities A, B, Cln, Eu
Propagation Separation of new bulbs (seed)
Use 1.1; 2.5; 3.4; 5.1; 6.2

Nicotiana spp. Tobacco plant. Solanaceae.
Facilities Cln
Propagation Seed
Use 10

Nuphar spp. Yellow water lily. Nymphaceae.
Facilities Cpn
Propagation Division
Use 2.5; 3.4

Nymphaea spp. White water lily. Nymphaceae.
Facilities Cpn
Propagation Division (seed)
Use 2.5; 3.4

Opuntia spp.[18,72] Various common names. Cactaceae.
Facilities A, B, Eh
Propagation Stem cuttings
Use 2.5

Ornithogalum virens. Star of Bethlehem. Liliaceae.
Facilities Cln, Ef
Propagation Separation of new bulbs
Use 3.7

Oxalis acetosella. Wood sorrel and other species. Oxalidaceae.
Sleep movements.
Facilities Cln
Propagation Seed/division of roots or offsets
Use 1.4

Paeonia spp. Peony. Ranunculaceae or Paeoniaceae.
Facilities Cln
Propagation Division of roots
Use 3.6; 3.7; 6.1

Papaver spp. Poppy. Papaveraceae.
Facilities Cln
Propagation Seed/division of roots in the perennial species
Use 7.3; 9.4

Pastinacea sativa. Parsnip. Umbelliferae.
Facilities Cln
Propagation Seed
Use 2.5; 9.1

Pelargonium spp. and cultivars.[61] Geranium. Geraniaceae.
Facilities A, B, Ef. Cln in summer. D.
Propagation Stem cuttings/seed
Use 1.5; 1.7; 2.7; 5.2; 7.4; 10

Phaseolus radiatus. Mung bean. Leguminosae.
Facilities B, Ef
Propagation Seed
Use 1.2; 1.4

Phlox spp. Polemoniaceae.
Facilities Cln, Eu
Propagation Seed/division of roots/root cuttings
Use 10

Physalis franchetii. Chinese lantern. Solanaceae.
Facilities Cln
Propagation Seed/division of roots
Use 7.3

Picea spp. Spruce. Gymnospermae, Pinaceae.
Facilities Cln
Propagation Seed
Use 8.3

Pilea microphylla. Pistol plant. Urticaceae.
Explosive buds which discharge pollen.
Facilities Eh
Propagation Stem cuttings
Use 4.2

Pilea cadierei.[73] Aluminium plant. Urticaceae.
Facilities A, B, Eh
Propagation Stem cuttings
Use 7.4

Pinguicula vulgaris. Butterwort. Lentibulariaceae.
Insectivorous plant.
Facilities Cmn, Ef
Propagation Seed/division
Use 1.6; 2.5

Pinus spp. Fir, deal, pine. Gymnospermae, Pinaceae.
Facilities Cla, Csa
Propagation Seed
Use 2.2; 2.7; 3.2; 3.3; 3.4; 3.5; 8.3

Pisum sativum. Garden pea. Leguminosae.
Facilities Cln, D
Propagation Seed
Use 1.1; 1.4; 1.8; 1.10; 2.5; 4.1; 9.1

Pleione spp. Ground orchid. Orchidaceae.
Facilities Ef. Can be grown outdoors in Cln. D.
Propagation Division of pseudobulbs
Use 2.6; 5.1

Primula spp. and cultivars.[60] Auricula, cowslip, primrose, polyanthus. Primulaceae.
Facilities Cln, Ef
Propagation Seed/division
Use 1.1; 4.2; 6.4; 10

Pteris spp. Bracken fern. Pteridophyta, polypodiaceae.
Facilities Cla
Propagation Division/spores
Use 2.2; 2.7; 3.2; 3.5; 5.1; 7.3; 7.4

Pyrus communis.[84] Pear. Rosaceae.
Facilities Cln
Propagation Grafting
Use 5.2

Quercus spp. Oak. Fagaceae.
Facilities Cln
Propagation Seed (grafting)
Use 2.7; 8.3

Ranunculus spp. Buttercup. Ranunculaceae.
Facilities Cln
Propagation Seed/division
Use 2.3; 2.4; 2.5; 2.6; 3.1; 3.2; 5.1; 6.1

Raphanus sativus. Radish. Cruciferae.
Facilities Cln
Propagation Seed
Use 2.5; 4.2; 6.4; 9.1

Rhodanthe manglesii (syn. *Helipterum*). Everlasting flower. Compositae.
Facilities Cln
Propagation Seed
Use 7.3

Rhoeo discolor (syn. *Tradescantia*)[73] Commelinaceae.
Facilities Eh
Propagation Stem cuttings
Use 3.7; 6.2

Ricinus communis. Castor oil plant. Euphorbiaceae.
Facilities A, B, Eh
Propagation Seed
Use 1.9; 1.10

Rosa spp. and cultivars.[79] Rosaceae.
Facilities Cln, Eu
Propagation Grafting/stem cuttings/seed
Use 2.7; 3.2; 5.2

Rubus spp. Blackberry. Rosaceae.
Facilities Cln
Propagation Seed/separation of new plants from rooted stolons
Use 2.5; 2.7; 5.1; 9.1

Ruscus aculeatus. Butchers broom. Liliaceae.
Facilities Cln
Propagation Division
Use 2.5

Saintpaulia ionantha.[78] African violet. Gesneriaceae.
Facilities B, Eh
Propagation Leaf cuttings/seed
Use 5.2

Salix spp. Willow, osier. Salicaceae.
Facilities Cln
Propagation Stem cuttings
Use 4.2; 8.2; 8.3

Sambucus spp. Elder. Caprifoliaceae.
Facilities Cln
Propagation Stem cuttings
Use 3.2

Sanseveria spp.[73] Bowstring hemp. Liliaceae.
Facilities A, B, Eh
Propagation Division
Use 7.4

Sedum spp. Stonecrop. Crassulaceae.
Facilities Cln, Ef
Propagation Stem cuttings/division (seed)
Use 2.5; 5.1; 5.2

Selaginella spp.[86] Tree club moss. Pteridophyta, Selaginellaceae.
Facilities B, Eh
Propagation Stem cuttings
Use 2.2; 2.7; 3.5

Sempervivum spp.[73] House leek. Crassulaceae.
Hardy succulent-leaved perennials.
Facilities A, B, Eu, Cln
Propagation Offsets
Use 5.1

Solanum tuberosum. Potato. Solanaceae.
Facilities Cln
Propagation Tubers
Use 1.9; 1.10; 2.5; 5.1; 9.1; 9.4

Statice vulgare (syn. *Limonium*). Sea lavender. Plumbaginaceae.
An 'everlasting' flower.
Facilities Cln
Propagation Seed
Use 7.3

Stellaria media. Chickweed. Caryophyllaceae.
Facilities Cln
Propagation Seed
Use 1.1; 3.7

Symphoricarpus albus. Snowberry tree. Caprifoliaceae.
Facilities Cln
Propagation Stem cuttings/division of new plants from rooted suckers
Use 1.3

Tagetes spp. African and French marigold. Compositae.
Facilities Cln
Propagation Seed
Use 10

Tilia spp. Lime. Tiliaceae.
Facilities Cln
Propagation Seed
Use 3.2; 8.3

Tolmiea menziesii. Pick-a-back plant. Saxifragaceae.
Facilities A, B, Cln, Eu
Propagation Division
Use 7.4

Tradescantia spp.[73] Spiderwort. Commelinaceae.
Facilities A, B, Eh. Some species are hardy Cln
Propagation Stem cuttings/division
Use 1.3; 2.6; 3.7; 5.2; 6.2; 7.4; 10

Trifolium repens. Clover. Leguminosae.
Facilities Cln
Propagation Division/separation of rooted plants on runners/seed
Use 1.4; 9.2

Triticum spp. Wheat. Graminae.
Facilities Cln, D
Propagation Seed
Use 1.6; 1.10; 2.3; 2.6; 7.3; 9.2

Tropaeolum majus. Nasturtium and other spp.
Tropaeolaceae.
Facilities Cln
Propagation Seed/division
Use 3.2; 10

Typha spp. Reed mace. Typhaceae.
Facilities Cpn
Propagation Division
Use 3.3; 7.3

Vallisneria spiralis.[62] Eel grass.
Hydrocharitaceae.
Aquarium plant.
Facilities A, B, Cpn
Propagation Division/runners
Use 2.5; 5.1

Vicia faba. Broad bean. Leguminosae.
Facilities Cln, for some work B
Propagation Seed
Use 1.1; 1.2; 1.10; 2.3; 2.4; 3.1; 3.7; 4.1; 9.1

Vitis spp. Grape vine. Vitaceae.
Facilities Cln, Ef
Propagation Stem cuttings/layering/seed
Use 2.5; 3.3; 9.1

Zea mays. Maize. Graminae.
Facilities Cln, D
Propagation Seed
Use 1.1; 1.4; 1.6; 1.9; 1.10; 2.3; 2.4; 3.2; 7.3; 9.1

Zebrina pendula (syn. *Tradescantia zebrina*).[73]
Wandering jew. Commelinaceae.
Facilities A, B, Ef
Propagation Stem cuttings
Use 1.3; 2.6; 3.7; 5.2; 6.2; 7.4; 10

Chapter 1 Using the classroom or laboratory

Only a small number of plants will grow satisfactorily in the classroom or laboratory without help. Amongst these are the commonly grown decorative pot-plants such as the geranium (*Pelargonium*), the ivy (*Hedera*), the painted leaf plant (*Coleus*), the spider plant (*Chlorophytum*) the spiderwort (*Tradescantia*) and the wandering jew (*Zebrina pendula*) (Fig. 1). These plants survive seemingly impossible adverse conditions and often, in consequence, look rather sickly.

Many of the signs indicative of poor growth given on page 40 will be only too evident. Diagnosis of the possible causes of failure will reveal the inadequacies of the surroundings.

It must be admitted that classrooms and laboratories are not the most ideal places in which to grow plants. The following factors, alone or in combination, prevent healthy growth for any length of time for many species.

1 Light—intensity and quantity usually inadequate, especially in winter. Illumination often unilateral from side windows.
2 Temperature—often far too high especially in the daytime and, if the plants are on windowsills, very low temperatures may occur in winter. Temperatures may fluctuate widely and rapidly due to draughts or timed central heating systems.
3 Humidity—too low for the majority of plants.
4 Watering—the school atmosphere tends to be dry. This means that frequent watering is usually necessary, which can be a problem at weekends and holidays.
5 Contamination—dust and chemical fumes are inevitably present. Spraying the foliage to clean it can be difficult if not impossible.

In addition to these undesirable environmental factors, storage for long term class investigations within the laboratory may prove difficult. These then are the major difficulties to be overcome and it is possible to do this with quite simple equipment.

Using the equipment described in this chapter not only can those plants already mentioned be grown much better but a greater range can be kept. Furthermore when some form of enclosure is used, be

Figure 1 Pot plants suitable for schools: a, Coleus; *b,* Sanseveria; *c,* Hedera; *d,* Begonia; *e,* Pelargonium; *f,* a cactus; *g,* Zebrina; *h,* Chlorophytum.

it a glass terrarium or a propagating case, and especially if it is heated, then propagation by cuttings or seed is much easier.

Accordingly this chapter is divided into two sections: the classroom or laboratory without any additional facilities and the laboratory with some facilities. With the exception of the terrarium and bottle garden these facilities are not usually suitable for use in classrooms.

A CLASSROOM/LABORATORY WITH NO ADDITIONAL FACILITIES

Plants are best grown on window ledges where they receive good light but may need to be protected from frost in winter. The pots should be stood in suitable trays or saucers. In these a thin layer of gravel or small pebbles can be placed. Water can then be poured in till its surface is just below the top of the gravel or pebble layer. The humidity around the plants is thus increased as the water evaporates whilst the plant pot is not standing with its base in water.

Short-term germination investigations, growing of bulbs and other indoor plants, production of root tips and propagation by seed or cuttings not requiring heat can all be carried out.

Figure 2 A terrarium made from a plastic aquarium. Planted with mosses and ferns a woodland habitat is simulated. Approximate dimensions: length 30cm (12in), width 20cm (8in), height 20 cm (8in).

Figure 3 Bottle gardens in a variety of containers, from left to right: Saintpaulia *in a fish bowl; mosses and ferns in a sweet-jar and in a 2lb storage jar. It is easy to place plants in bottles with wide necks but water loss can be a problem, unless the top is covered.*

B LABORATORY WITH SOME FACILITIES

The following equipment can be used to minimise or overcome the problems mentioned above.

1. Terrarium

Terrariums are available commercially with metal frames and slanting hinged glass fronts. Alternatively any large glass container or aquarium is suitable, with glass or plastic top cover (see Fig. 2). A layer at the bottom of equal parts of coarse washed gravel and coarse charcoal is added followed by the material in which the plants grow.

A desert habitat may be simulated in which cacti can be grown if the material has a high proportion of sand. Using a soil-less medium or John Innes Compost No. 1 a woodland habitat can be simulated particularly suitable for the culture of liverworts, mosses and small ferns. If the bottom layer of gravel, charcoal and medium is piled up to one end, which is also raised by placing a small piece of wood under the terrarium, then a small amount of water can be poured into the lower end and a semi-aquatic habitat simulated.

It is advisable to cover terrariums closely with a sheet of glass, protecting the edges with adhesive plaster or plastic. If this is done the loss of water will be very small and they will require little attention, after being set up in a bright place, other than occasional watering and the removal of dead leaves. If the growth of the plants is slow they need not be disturbed for a long time.

2. Bottle gardens

These are prepared in a manner similar to the terrarium.[5 7] It is best to use a soil-less medium. For planting, long wooden forceps are helpful. Water is added through glass or flexible plastic tubing, so that it does not touch the sides. The top can be left open (see Fig. 3).

Mosses, small ferns and the following flowering plants are suitable; aluminium plant (*Pilea cadierei*), *Begonia*, *Bryophyllum*, *Gynura*, ivy (*Hedera*), painted leaf plant (*Coleus*), spider plant (*Chorophytum*), spiderwort (*Tradescantia*) and the wandering jew (*Zebrina*).

3. Polythene enclosures

These can be made using a wooden, plastic or metal tray partly filled with gravel or sand as a base, with a covering of polythene held by a light framework of galvanised wire or canes (see Fig. 4). Alternatively the entire base and spacing framework could be enclosed in a large plastic bag. Plants are best grown in pots which are placed on the gravel or sand. This allows some water to be left in the base, so that a high humidity can be provided without the medium or compost in the pot becoming water logged.

4. Propagating cases

These are, in effect, small self-contained glasshouses consisting of a plastic, resin reinforced glass fibre or wooden tray with a glass or plastic cover and electric heating at the base (base heating). The sides may be raised or a domed plastic lid fitted to increase the headroom. They are suitable for use in the laboratory and greenhouse (see Fig. 5).

Several commercial models of different sizes are available with electric base heating either by warming cables or an integral heater base. Some of the deeper ones can be fitted with similar cables around the sides to heat the air inside (space heating). Full instructions are provided with all these models.

Simple propagating cases can be constructed. A waterproof base of wood, resin reinforced glass fibre or metal is required. Wooden bases may be waterproofed with a lining of heavy gauge polythene sheet. The dimensions are not critical and should be such as to conveniently accommodate seed-trays or pans. (The dimensions of a standard full size seed tray are approximately 23 x 38 cm (9 x 15 in). They are usually about 5.5 cm (2.25 in) deep.) Heated propagating trays which make an ideal base are available commercially. Over the base a galvanised wire frame covered with transparent plastic sheet should be added.

Base heating may be provided by embedding an electric warming cable halfway down a layer of gravel or sand about 10–15 cm deep (4–6 in).[45,48] A loading of 86–130 W/m^2 (8–12 W/ft^2) would be suitable. The cable should be laid in even loops, a 75 watt/25 ft cable being suitable for a base area of 0.5–0.75 m^2 (5–8 ft^2). Low voltage rather than mains cable is recommended for safety. The cable could be connected to a suitable thermostat embedded in the gravel or sand at a depth of about 2 cm (0.75 in).

Seeds, cuttings or small plants can be grown in pots, boxes, plastic trays, or seed-pans and

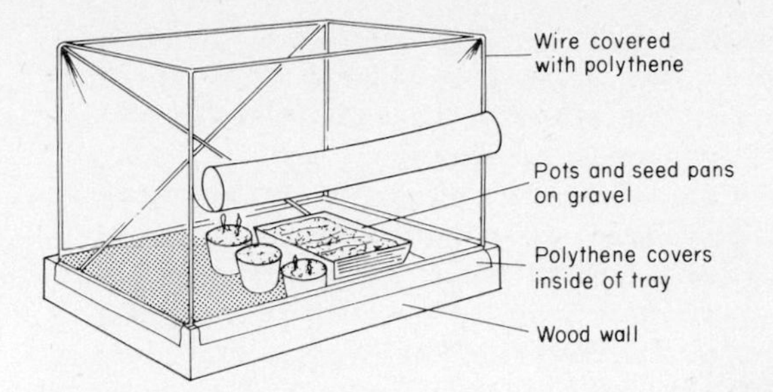

Figure 4 A simple polythene enclosure.

Figure 5 A commercial propagating case. Base of glass fibre/resin with electric heating and removable cover of transparent plastic. Strip lighting may be attached to the base. Approximate dimensions: width 93cm (37in), depth 70cm (28in), height 47cm (19in).

maintained in the warm humid atmosphere.

Space-heating could be provided in a manner described for the frames on page 28 and the loading calculated to provide a maximum temperature rise of 5°C (9°F), the thermostat being set at 18–21°C (65–70°F).

5. The Wardian case

This is a miniature glasshouse designed to provide a fully controlled environment.[59] It can be built out from existing windows so that access is through the window or it may be built inside the laboratory in a similar manner to a fume chamber. When built outside, the internal opening should be made as large as possible to give ease of access and the window must either be sliding or be hinged to open into the room. A cross section of a suitable arrangement is shown in Fig. 6.

The case should have a waterproof tray either of metal, plastic or resin reinforced glass fibre covering the whole area of the base. There should be provision for base heating by either low-temperature heating tubes or an insulated soil warming cable embedded in gravel, and thermostatically controlled space heating and lighting controlled by a time switch. If it is built on the side of the building exposed to the sun then electric fan ventilation will be necessary otherwise manually opening ventilators will be satisfactory.

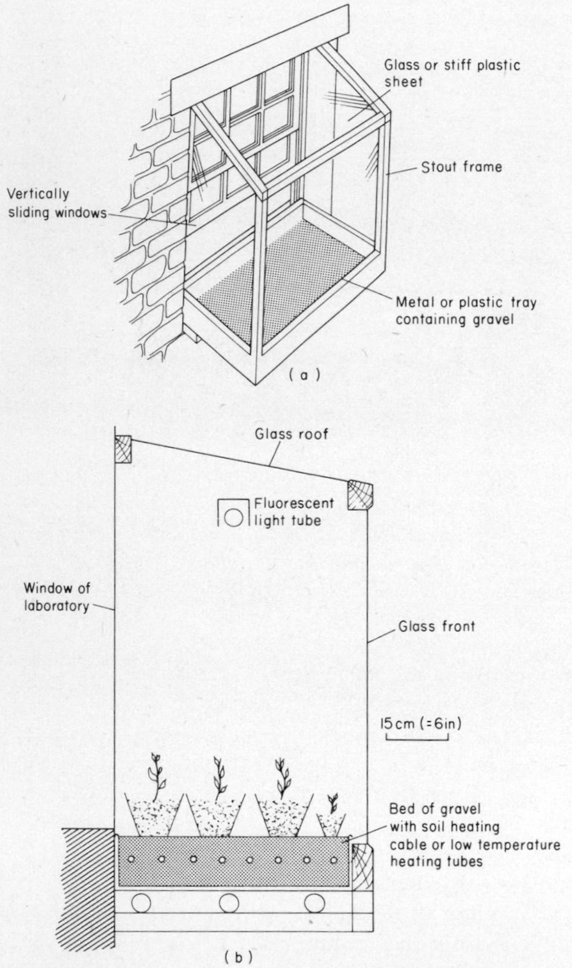

The insulated soil-warming cable should be of the low voltage type embedded halfway down in a bed of gravel about 10–15 cm deep (4–6 in) as for the propagating frame. A loading of from 108–162 W/m^2 (10–15 W/ft^2) is satisfactory. If chicken wire mesh is laid just above the cable the heat is spread more effectively. For economic operation the cable should be connected to a suitable thermostat embedded in the gravel to a depth of about 2 cm (0.75 in).

For space-heating, tubular heaters are most satisfactory, but mineral-insulated, copper-sheathed warming cables could be used. The calculation for the wattage required is similar to that on page 34 and if the case is external then a minimum rise of 11°C (approx. 20°F) over that of the exterior will be required. A thermostat to control the space heating is essential. The lighting requirements are similar to those for glasshouses given on page 34.

Provision of a number of Wardian cases gives flexibility as each can be controlled at a different temperature, humidity and level of lighting. Wardian cases installed outside suitable corridor windows are an added attraction as long-term investigations can be observed without the need for access to laboratories and interesting plants can be seen by all members of the school.

Glass or stiff plastic sheet
Stout frame
Vertically sliding windows
Metal or plastic tray containing gravel
(a)
Glass roof
Fluorescent light tube
Window of laboratory
Glass front
15 cm (=6 in)
Bed of gravel with soil heating cable or low temperature heating tubes
(b)

Figure 6 The Wardian case. Length and height are determined by the size of the window. (a) A simple Wardian case. (b) Cross-section of a Wardian case.

6. Plant trolley

This is a mobile unit similar to a laboratory trolley mounted on castors incorporating a substantial plant storage area with shelves, bearers and, if necessary, trellis work. Automatic watering, and possibly a propagating case can be incorporated. In it a wide variety of plants, including climbers can be grown.[59] A few designs are available commercially. However they can be easily constructed using slotted metal angle strip or wood. If they are designed to be narrower than the door openings, the resulting mobility between rooms is a great advantage.

Potted plants are stood on coarse sand which completely fills waterproof metal or resin reinforced glass fibre trays. The sand can be kept moist by a drip feed controlled by a screw clip on a plastic tube (Fig. 7a) or by an arrangement similar to a poultry drinking trough (Fig. 7b). This use of capillary

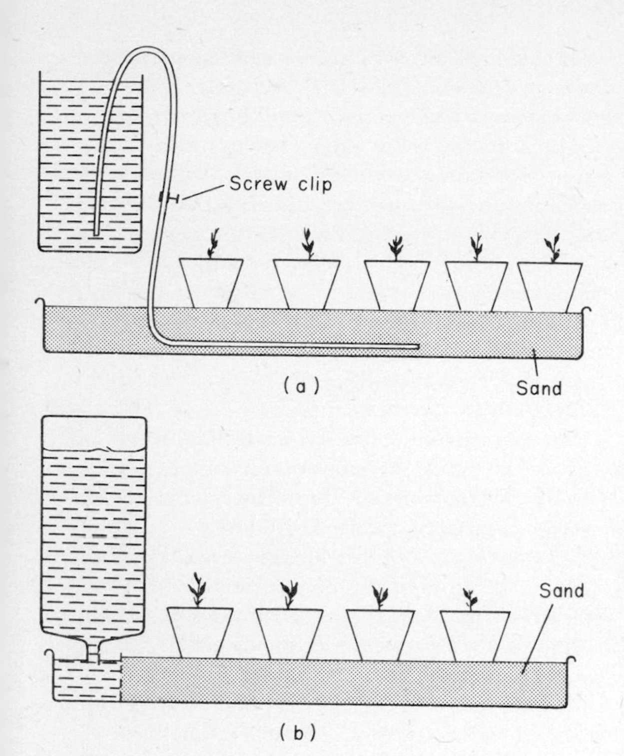

Figure 7 Capillary trays: (a) with drip feed; (b) with constant level feed.

Figure 8 A movable plant trolley made with a framework of slotted metal angle strips. The two compartments have fluorescent lighting and fan ventilation. The upper can be removed and used separately as a lamp frame. Sprayers, labels and other equipment can be stored on the lower shelf. Approximate dimensions: width 115cm (46in), depth 65cm (26in), height 165cm (66in).

watering overcomes the low humidity of the atmosphere. A standard commercial capillary tray using a one-gallon jar as a water reservoir provides suitable automatic watering for standing plants. This form of plant trolley could be more or less completely enclosed using plastic sheet or, less suitably, glass.

Electric lighting and heating can be incorporated if desired (see Fig. 8). Information on lighting requirements is given on page 24. Base heating similar to that for propagating cases would be suitable. The addition of these features would obviously limit mobility but the unit could be connected to the electricity at night and moved into its daytime display position early in the morning.

7. The lamp frame

One of the major deficiencies in the laboratory which affects normal plant growth is poor illumination, both in intensity and duration. The level of illumination can however be increased by suspending fluorescent tubes or lamps and/or incandescent lamps above the plants. A minimum intensity of about 8600 lux (800 foot candles or lumens/ft^2) is normally necessary, assuming a suitable range of wavelengths is available, for most plants to balance the loss of products in respiration by those produced during photosynthesis. However at this low intensity the critical balance of the reactions in the flowering mechanism in many plants may be interfered with giving abnormal responses.[41] For the normal growth of most plants suitable for school use intensities of 8600–21520 lux (800–2000 fc) will be adequate.[42] In lamp frames for schools an ideal minimum intensity would be 10760 lux (1000 fc).

The simplest lamp frame is illustrated in Fig. 9a and consists of a large concave reflector made of a number of metal or cardboard segments, which can be conveniently separated for storage, mounted above

Figure 9 Lamp frames. (a) A very simple way to increase the available light for a few plants. An 'umbrella' made of 'petals' of white cardboard over a mercury vapour MBTh lamp. The 'umbrella' can be collapsed for storage. Diameter approximately 75cm (30in).

(b) A free standing lamp frame covered with polythene film. Three 30 watt fluorescent strip lamps can be fitted. Approximate dimensions: width 115cm (46in), depth 60cm (24in), overall height 140cm (56in), height of plant cabinet 65cm (26in). See also Figure 8.

a suitable high intensity fluorescent lamp (the usual rating of which is 400 watts) the whole being supported on a large retort stand or tripod legs.

Larger lamp frames can be constructed using metal strips or wood into which a battery of fluorescent tubes and their control gear can be fitted (See Fig. 9b). 'Warm-white' tubes are very suitable as the light they emit has a relatively high red content. As these lamps have a low surface temperature the distance between lamp and plant can be small without risk of scorching the foliage. If it is desired to increase the amount of visible red light and also the infra-red radiation then incandescent lamps can be added with a wattage ratio of 1:3 incandescent light to fluorescent light.[42] However with these lamps local heating and scorching of the foliage can be a problem.

The larger lamp frames should be between 45–75 cm (18–30 in) high, this being adequate for most pot plants and seedlings in boxes. The tubes should be placed no more than 7.5 cm (3 in) apart to give maximum intensity. Ideally the lamps and plants should be arranged so that the distance between lamp and plant can be varied. In view of the range of plants required which vary both in the light intensity required and in height, it is most convenient to place the plants on suitable blocks to bring them near to the lamps. A possible range of spacing of 15–60 cm (6–24 in) between plant growing surfaces and light source surfaces would be ideal to accommodate both pot plants and seedlings in boxes or trays. A light intensity of approximately 10760 lux (1000 fc) is obtained 15 cm (6 in) below the centre of two 40 watt tubes spaced about 2.5 cm (1 in) apart. Over 21520 lux (2000 fc) is obtained in the same position below six 40 watt tubes similarly spaced. Clearly the intensities will vary depending on the type of tubes and reflectors used.

Further information on lamp spacing and resultant intensities may be found in references 42 and 45.

8. The plant cabinet

The logical extension from the lamp frame is to enclose this within a cabinet to which electrical air circulation and heating equipment are added thus enabling control of the internal environment. A cabinet of this type was constructed and evaluated by the Educational Use of Living Organisms Project. The design and dimensions are shown in Fig. 10. It was made from sheet aluminium on a metal tube frame.

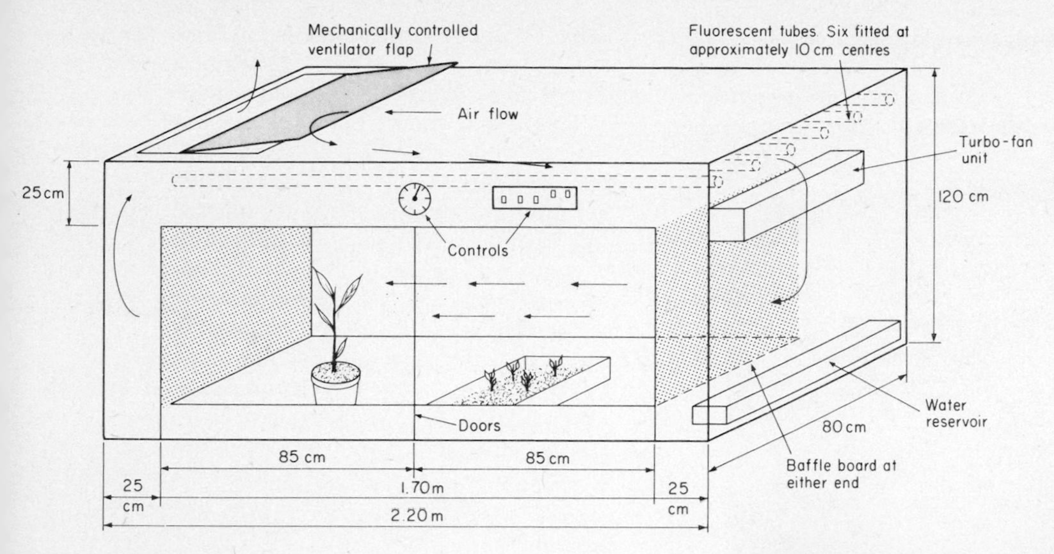

Figure 10 Plant cabinet.

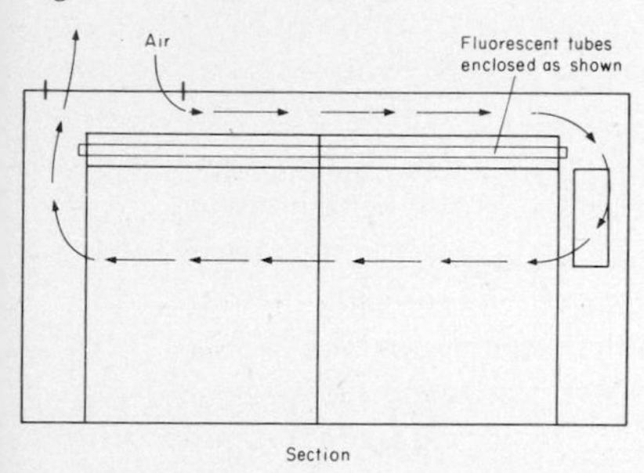

Within the cabinet, provision was made for racking to accommo te plants of varying heights. An electric heater with a small integral fan unit was used. The internal temperature was thermostatically controlled and the ventilator flap controlled mechanically by an adjustable automatic greenhouse ventilator unit. The cost in 1972 was about £20.

During evaluation, internal temperatures and light intensities were measured, air movement estimated and the suitability of the cabinet for housing a variety of living organisms in schools assessed. With an ambient temperature of 20°C (68°F) trials showed that the internal temperature reached 28°C (82°F). This was due to the heat generated by the lamps and associated starters which were enclosed for electrical safety. This enclosure also reduced the light intensity considerably. Thus it would be best to leave the fluorescent tubes exposed, enclosing only their ends, the sockets and the starters.

A variety of plants, including algae, commonly grown in schools were successfully maintained. It was found that cultures of animals tolerant to relatively high humidities such as *Drosophila* and *Tribolium* could be kept with these plants. Alternatively the cabinet could be used for keeping animal stock alone, such as small mammals.

The construction of a similar cabinet from plywood on a wooden frame could form an excellent project for a school. The critical dimensions are those of the length of the fluorescent tubes and of the trays or boxes used (the dimensions of full size seed trays are given on page 21). Other dimensions are not critical and may be varied to suit the individual situation. Plants of different heights can be accommodated by using movable shelves. Electrical safety is of particular importance and all electrical components and metal parts must be adequately earthed and suitably protected from water.

Chapter 2 Keeping plants outdoors

Figure 11 Plants can be grown outdoors in a variety of containers. (a) Daffodils, anemones and other herbaceous plants in half wooden barrels about 65cm (26in) top diameter; aquatic plants in a plastic tray; approximate dimensions: length 55cm (22in), width 42cm (17in) depth 15cm (6in); dwarfed Ginkgo *in large plastic flower pot 27cm (11in) top diameter.*

(b) Alpine plants in a plastic tray, dimensions: length 60cm (24in), width 30cm (12in), depth 10cm (4in).

Suitable plants can be grown outdoors by most schools. A few will have to make do with window boxes, others more fortunate will have garden plots.

There is much that can be done even on unpromising sites. Window-boxes or free standing wooden plant troughs can easily be made in which bulbs and, later, annual flowering plants can be grown (see Fig. 11). Soil beds can be built up using a retaining wall of peat blocks or old well weathered railway sleepers. A wide range of plants can be grown in these beds. In this way children in schools in the centre of cities can enjoy the experience of outdoor gardening.

If some form of protection against the extremes of climate is given then the range of plants that can be grown out of doors is considerable. The sections of this chapter after the first are concerned therefore with the necessary facilities to achieve protection.

C. OUTDOORS: UNPROTECTED

The garden

The value of a school garden depends on size, aspect, soil type, climatic conditions, pollution and, in some schools, the incidence of vandalism outside school hours.

One natural feature which can be partially controlled is the soil type. Dry sandy soils can be made to retain water by the addition of organic material in the form of manure, composts, peat, spent hops and a variety of textile wastes. Water-logged acid soils on a clay base can be improved by sub-soiling, drainage, the use of lime and incorporation of organic material. Many plants will grow equally well in a variety of soil types and in a wide range of acidity/alkalinity (pH) values.

Highly alkaline soils (those with a high pH) can seldom be made acid. Some measure of success in growing lime-hating plants (calcifuges) such as the heaths (*Erica*) can be achieved using peat, applying solutions of aluminium sulphate or chelated iron compounds (sequestrated iron) and fertilising with sulphate of ammonia.

The most useful soil for a school garden is a good

fibrous loam with a neutral reaction. Few plants will fail to grow in this type of soil and for the more extreme lime-loving plants (calcicoles) the incorporation of old mortar rubble or limestone will provide the ideal conditions.

D. OUTDOORS: PROTECTED

1. Cloches

Cloches can be used to cover plants in the garden when growing singly or in rows. They can be moved easily from plant to plant or row to row as needs change.

A wide variety is available commercially ranging from tent cloches consisting of two pieces of glass or rigid plastic sheet held together by a clip, barn cloches with four pieces held together by a wire structure to give a greater height to cloches made of thin flexible plastic sheeting supported by a wire frame (see Fig. 12).

Whilst they give protection from wind, excessive drying up, too much rain, birds and snow they do not afford any frost protection to crops under them.

As far as possible, allowing for maximum lighting, the rows of plants to be protected by cloches should be at right angles to the wind direction. The ends of cloches or lines of them, especially the light plastic varieties, must be closed to prevent a tunnel for the wind being formed.

Figure 12 Two types of glass cloches. In the upper photograph a tomato crop is growing in a trench over which the cloche is placed.

2. Frames

Using frames is an ideal way of keeping a large collection of alpines, bulbs, hardy material for genetics, e.g. *Primula,* antirrhinum and radish, collections of cuttings of shrubs and plants needed later for investigations in the laboratory, e.g. runner beans in pots.

Frames can be made of brick, cement blocks, wood or asbestos to give a depth varying from 30–75 cm (12–30 in) or more. They are covered by glazed lids or lights. These are usually wooden framed and may have window bars to support the panes of glass which are fixed by embedding in putty. Alternatively, in the Dutch light, the wooden frame is grooved to hold one large piece of glass or rigid plastic (see Fig. 13). Commercially made frames are also available with metal framing into which standard sized sheets of glass or rigid plastic can be slotted or clipped as the sides and lid, thus allowing much more

Figure 13 (a) Dutch light frame with wooden framework.

Figure 13 (b) Commercial model with metal framework. Note the air heating cables and rod thermostat.

light in. Horticultural or 24 oz. glass is 0.3 cm thick and available in the following sizes: 40.6 x 61.0 cm (16 x 24 in) 45.7 x 61.0 cm (18 x 24 in) and 61.0 x 61.0 cm (24 x 24 in).

Unheated frames may be used for keeping many hardy and half-hardy plants. Plants grown in clay pots, which are plunged to their rim in a layer of peat or ash, are readily available for transfer to the laboratory and their upkeep is much simpler than in glasshouses. Plunging the plant pots provides a much more stable environment for the roots since rapid water and temperature changes do not occur. Covering by frame lights or Dutch lights protects plants from the weather and gives control over their watering. In very cold weather it may be necessary to cover the frames, in particular the lights, with sacking to minimise the effect of frost.

Heating the soil or gravel base and the air within the frame greatly increases its versatility.[45,48]

The soil warmed frame is ideal for encouraging plants such as lettuce and brassicas which are later to be planted out. Heating cables, preferably of the low voltage type, should be buried directly in the soil at a depth of about 15–20 cm (6–8 in) and evenly spaced to give a loading of about 65 W/m^2 (6 W/ft^2). With the low voltage cabling the necessary transformer may be stood directly on the soil in the corner of the frame or attached to a side. Soil cultivation must be done with care once the cables are installed and only when the electricity is switched off. It is unnecessary to leave the heating cable switched on all day unless very low temperatures occur and control by the 'dossage' method is suitable. This can be done either manually or by a time-switch. In the warmer parts of the country a running time of 10 hours per night will give a good soil temperature, in colder parts 12–14 hours may be necessary. Should the frame be used to raise plants in pots then the depth of soil above the heating cable should be about 10 cm (4 in) and the pots plunged to a depth of about 5 cm (2 in) in the soil.

Propagating beds in frames can be made on any ground free of water-logging or on concrete or gravel. The heating cables, again preferably of the low voltage type, should be laid on a layer of coarse sand about 5 cm (2 in) deep and then covered with another similar layer. A loading of 86 W/m^2 (8 W/ft^2) is suitable and the distance between the cable loops should be between 5–10 cm (2–4 in). Plants can be raised in pots or boxes stood on top of the sand layer.

Space-heating the frame is achieved by fixing the heating cable to its side with suitable insulators. It is best controlled by a rod-type or other suitable thermostat positioned between 10–15 cm (4–6 in) above the surface of the soil.

For protection against frost in all districts, except those which experience extremely low temperatures, a loading for the space-heating which gives a temperature rise above that of the exterior of about 11°C (approximately 20°F) is suitable. Estimation of the total wattage required can be made after measuring the total area of glass in the frame as follows:

Area of glass in m^2 x 11.6 x temperature rise in °C required = total wattage required for space-heating 1

(Area of glass in ft^2 x 0.6 x temperature rise in °F required = total wattage required for space-heating).

Thus if 1 m^2 = 10.8 ft^2 and 1°C rise = 1.8°F rise then assuming a frame with a light of 1 m^2 (about 40 x 40 in) and a desired temperature rise of 11°C (19.8°F)

1 x 11.6 x 11 = 128 watts

(10.8 x 0.6 x 19.8 = 128 watts)

This is the minimum wattage required to give protection against frost for a frame and if this is mainly of glass then it would be wise to increase it by some 40–50% (to 179–192 watts). To allow the air in the frame to heat up in sufficient time to prevent frost damage the thermostat should be set to about 2°C. A frost protected frame will enable plants such as geraniums and fuchsias to be over-wintered in addition to housing a few other pot plants. It can also be used to harden off plants from the greenhouse before they are planted out in the open.

If the frame is needed to keep a range of plants other than those for which only protection against frost is required then a higher internal temperature rise is necessary coupled with soil heating. For soil warming the heating cable preferably of the low voltage type, should be buried under a layer of sand about 7.5 cm (3 in) deep. A loading of 65 W/m^2 (6 W/ft^2) is suitable. The sand provides a warm bed on which pots or boxes can be placed. Clay pots are best plunged into a layer of peat about 10 cm (4 in) deep on the sand. Cuttings may be rooted in a layer of soil-less medium or suitable compost about the same depth. The soil warming can be manually controlled or time switched to heat for about 18 hours per day, say from 4 pm to 10 am the following day, to maintain a soil temperature of about 18°C (65°F). The loading for the space heating can be calculated using equation 1 (page 28) assuming a minimum temperature rise of 13.5°C (25°F) which should maintain an internal air temperature of 7°C (45°F). In very exposed areas the desired temperature rise should be higher.

3. The Bird-proof cage

This is a necessity for any investigations where seeds have to be collected. This should be large enough for a person to move about inside. Such crops as peas, wheat, barley and maize may be grown in large pots or in the open ground and the seeds required for genetics work may be collected. A stout wooden structure, covered by half-inch mesh wire netting is adequate. A locked entrance will help keep out intruders (see Fig. 14).

Figure 14 A large bird-proof frame for growing soft fruit or small crops of cereals. The height is approximately 2.4m (8ft).

Chapter 3 The glasshouse

Figure 15 A well designed glasshouse with attached cold frame. Note the slatted shading, guttering and fall pipe. A glass to ground house may be better for school use and a sliding door is advisable.

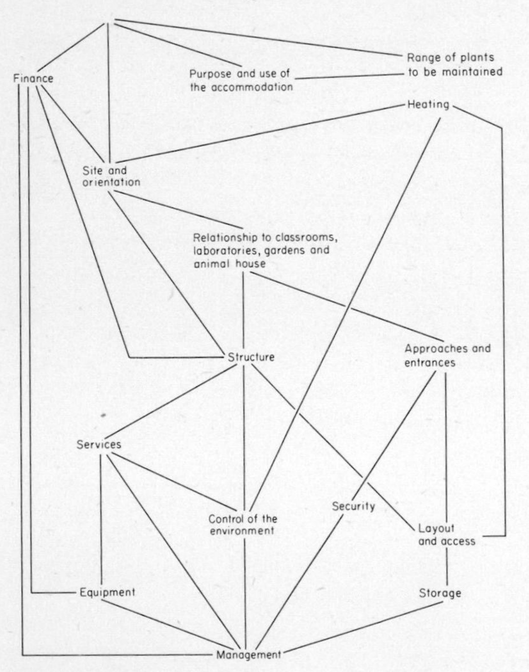

Figure 16 Glasshouse planning points and their interaction.

A fixed glasshouse is an essential adjunct to any biology department in that it enables a wide variety of plants to be maintained, and can be used to supplement the work in the laboratory, particularly for those investigations requiring a controlled environment (see Fig. 15).

All electrical equipment must be correctly installed, adequately safeguarded by the use of appropriately fused circuits, used correctly and satisfactorily maintained (see pages 33 and 36).

Whenever glasshouse accommodation is being planned some fourteen points need to be considered. The interaction of these is shown in Fig. 16 and they are considered in detail below.

PURPOSE

It is suggested that the purpose of a school glasshouse should be:

1 to provide a supply of plant material for use within the school.
2 to enable a variety of plant species particularly suitable for school use to be maintained in good health.
3 to provide facilities for pupils to observe plants without unduly disturbing them.
4 to provide facilities where the correct procedures relating to the maintenance of plants can be seen, practised and the related biological principles studied.
5 to provide facilities for investigations involving plant material which require a controlled environment.

(A similar approach to the design of animal accommodation is to be found in the companion publication 'Animal Accommodation for Schools'.)

Essentially the school glasshouse has two main functions. Firstly that of housing a collection of plants and secondly as an area in which investigatory work can be carried out. Since both of these functions necessitate the use of soil or bench space in a totally different way they are not necessarily compatible. Furthermore the facilities required are different.

For housing and maintaining an adequate collection, all the services and equipment described in this chapter will be necessary at some time. For investigatory work propagating equipment such as capillary trays, mist propagation and soil warming will be necessary. For the temporary attachment of electrical apparatus, solid instead of slatted staging can be advantageous.

Careful consideration must therefore be given as to the purpose and desired function of the glasshouse before any particular design or site is chosen.

FINANCE

The cost of domestic greenhouses in 1973 with glass to ground-level, which, in general, have maximum widths of about 4 m (12 ft) was between £11–22/m^2 floor (£1–2/ft^2). The smallest width available in commercial glasshouses with glass to ground is usually about 5 m (15 ft) and the cost per unit floor area for these is approximately half that of a domestic house. However since the size of the smallest commercial glasshouse is much greater than that of a domestic glasshouse, the total cost of a commercial house, excluding fittings, is relatively high. Care should be taken in choosing individually designed 'one off' glasshouses since the cost per unit area is usually higher than domestic types and the designs may often be unsatisfactory from the horticultural and management point of view especially in terms of ventilation.

To the initial cost of purchase must be added that of the equipment necessary. This will be approximately equal to the initial purchase cost for domestic greenhouses, somewhat less than this for commercial ones. If the house is some distance from water and electricity then the cost of installing these services can be considerable.

For effective use glasshouses with a minimum floor area of approximately 40 m^2 (430 ft^2) are recommended. In these not only can a demonstration collection be housed but space will be available for investigatory work. A minimum width of 5 m (15 ft) is desirable since this will allow both side-staging or beds and a centre bench together with adequate space for movement of pupils. Should finance be difficult it is better to dispense with staging and other fittings rather than choose too small a house.

To the cost of the house and equipment it may be necessary to add that of erection. Expenditure after the installation will consist of the recurrent costs of heating and lighting, of pest control, and of maintaining the house, the plants, the equipment and services. This is difficult to assess since there are many variables. A figure of at least 10% of the initial purchase cost should be allowed for.

RANGE OF PLANTS TO BE MAINTAINED–HEATING

During the warm summer months almost any plant can be kept in a glasshouse without additional heat.

The range of plants that can be permanently housed depends on the minimum winter temperatures in the glasshouse. To a certain extent this depends on the exposure of the site and the orientation of the glasshouse. It is also possible to partition off part or to use specialised equipment, such as propagating cases to provide areas with different temperatures. Provision of controlled space or air heating is all important and in this respect glasshouses are of three main types.

1. Unheated (Eu)

These are usually used as an alpine house to grow plants in pots or pans which require adequate ventilation but need to be kept dry throughout the winter. Side window and under bench ventilation are essential to maintain a buoyant atmosphere. In this type of glasshouse a collection of high alpine plants with close cushion habits, hairy leaves, waxy cuticles or plants which produce a large amount of farina on the leaves can be kept. Such plants must be watered by plunging the pot or pan in a dish of water as overhead watering washes off the farina and may promote rotting of the hairy leaves or the cushions.

2. Frost proofed (Ef)

Within this type the minimum temperature is kept just above freezing point, usually about 2°C (36°F). Thus the heating costs are relatively low.

This type is most useful for overwintering such plants as zonal pelargoniums (geraniums), fuchsias and many subjects which will not stand the wet and cold of an English winter. They are also adequate for growing on half-hardy annuals in the spring, before the danger of frost has passed.

3. Heated (Eh)

This type is undoubtedly the most satisfactory

solution to the maintenance and provision of a wide range of plant material. A minimum temperature of 7°C (45°F) is maintained. For school use it is rarely necessary for higher temperatures. However should these be required it is best to partition off the end of the glasshouse away from the door or to use a polythene enclosure as the high temperature space.

The numbers and variety of plants to be maintained whether they are to be pot-grown or grown inside beds of soil, their maximum size and habit of growth will all influence decisions as to the size of glasshouse necessary and its internal layout.

In the sections describing services, control of the environment and management this type of house is considered.

SITE, ORIENTATION AND RELATIONSHIP TO LABORATORIES, GARDENS AND ANIMAL HOUSES

The ideal site would be unshaded by buildings and trees, yet sheltered from the north and east to reduce heating costs and possible damage by wind, close to service points and conveniently located in relation to laboratories and gardens for ready access to collect specimens or for class use. In urban areas a rooftop site may be the only possibility, but here temperature control and low water pressure can be a problem. There may be local government regulations with regard to siting and any necessary planning permission must be obtained.

If a free choice of site is possible then an east to west orientation will give the best light in winter but temperature control may be a problem in the summer. In this orientation the door should be in the west end. A north to south orientation will give less light and may be best for small houses. The warmer end will face south and the door should be in the north end.

Animal houses should not be positioned close to glasshouses to avoid possible contamination from pesticides.

STRUCTURE, APPROACHES AND ENTRANCES, SECURITY

A detached span-roofed type with glass to ground is recommended. The framing should be metal, preferably aluminium, to minimise maintenance costs and loss of light. Those with wooden frames are less satisfactory.

The span-roofed house has vertical or inward sloping sides rising to the eaves. Their height should be at least 1.5 m (4.5 ft) preferably 2.0 m (6 ft). The pitched roof rises to a ridge, which should be about 0.6 m (2 ft) higher than the eaves. The pitch or slope of the roof must be steep enough to prevent snow lying which may cause damage.

The house should be glazed with standard size horticultural glass. If the panes are clipped to the framing they can be easily removed for replacement. Glass substitutes are, in general, not as satisfactory as glass although less fragile. They could be considered for glazing when vandalism is a problem. If this is so then at least it is advisable to use wired-glass for the roof to minimise possible damage to the plants and people from falling glass.

The foundations must be substantial and to a depth of at least 22.5 cm (9 in). The house must be firmly anchored at the corners to prevent movement and distortion, especially in exposed windy sites. On flat roofs it can be mounted on a dwarf wall of one or two brick courses with a few gaps to allow the entry of services and drainage. If the ground around the house is to be cultivated then the top of the foundations must be above ground level, so that the glass is protected from hoes and similar tools.

Simple lightweight plastic guttering should be fixed to the framing just below the eaves. This is not necessary with some aluminium framed houses which have an integral narrow channel. Small section guttering must be used so that a minimum of light is cut off. If convenient the downpipe could connect to a drain. Alternatively it should lead to a water butt or be directed to the lowest part of the site. The rainwater collected in the butt can be used for routine watering unless a shading wash has been applied (see page 34).

Sliding doors are essential; locks on the doors are desirable so that the house can be secured overnight and during weekends and holidays. The approaches should be visible from the laboratory.

If disabled children in wheelchairs are to use the house or if trolleys or wheelbarrows are used then ledges or sills at the door(s) must be avoided or be removable.

LAYOUT AND ACCESS

The internal layout will depend on the use to which the house is to be put. Inevitably some compromise will be necessary and it is better to plan for a multipurpose house in which a demonstration collection can be housed and some investigatory work can be carried out. The routine operations involved in propagation such as filling pots and trays or transferring seedlings can be carried out in the house but to do this considerable space is required. It may be necessary to do these jobs in the laboratory or on tabling out of doors.

If movable staging or benching is used this considerably increases flexibility. Below the staging, ferns, liverworts, mosses and other shade-loving plants can be grown either in pots or directly in side beds.

A continuous path enables easy access and movement by pupils. It is best made from concrete paving slabs so that it can be repositioned if necessary. The slabs should be placed with the rough face uppermost so that the surface is not slippery.

STORAGE

Secure storage is essential particularly for pesticides. It is best to keep these in a locked cupboard and many require storage in frost-free space.

Soils and composts should be stored in bins with twist lock lids. Those with wheeled bases are particularly useful being easy to move.

Tools and other equipment should be stored on suitable racks. It may be convenient and is economical on space to store bulky materials such as composts in a small shed.

SERVICES

Water and electricity are needed. The main controls should be placed near the door. All power points, switchboards and light fittings and other electrical apparatus must be waterproof.

Water

Copper piping is recommended with a main stopcock or taps on each outlet as required. The stopcocks or taps should have fittings to which hoses can be connected or be fitted with a short length of copper pipe to which hoses can be connected with adjustable circlips. The following outlets will be necessary:

1. tap at a height suitable for filling watering-cans
2. capillary trays or beds
3. mist propagation
4. overhead watering to staging
5. trickle line to side beds
6. spare for possible use as supply to other capillary trays.

A certain minimum water pressure is necessary for some equipment to work correctly, in particular mist propagation units and some capillary beds. It is advisable to have the pressure checked to see if it is sufficient before these are installed.

Electricity

Outlets for the following will be necessary:

1. main lighting
2. tubular heaters
3. extractor fan
4. air-circulation fan
5. low voltage soil warming cable for propagation bench
6. low voltage soil warming cable for soil in side beds
7. electric controller for mist propagation
8. electric fan in aspirated thermostat

9–10. at least two waterproof power sockets one at each end of the house for temporary attachment of additional electrical apparatus

The electric circuit to the greenhouse must be independent on a totally separate circuit so that a supply can be maintained at all times. The control panel should be placed near the door but in such a position that it does not cut off too much light. If the panel is boxed then it is advisable to have a small heater inside to prevent condensation.

CONTROL OF THE ENVIRONMENT

The temperature, lighting, air-movement and humidity are the major factors that need to be controlled within prescribed limits.

Temperature

Electric heating by tubular heaters is recommended unless the area of the house is above 1000 m^2 when oil or gas central heating is more economical. Fan heaters can be used but they are, in general, less reliable, although they do create essential air movement. Few schools are likely to use fan heaters with transparent plastic perforated tubes or ducts to ensure a more uniform distribution of warmed air, a practice increasing in commercial houses.

Tubular heaters are usually of 5 cm (2 in) diameter and these have an electric loading of 60 watts per 30 cm (1 ft) run of tube. They are best fixed singly or in tiers evenly around the walls of the house to ensure good heat distribution. The lowest tube should be about 10–15 cm (4–6 in) above ground level and spaced about the same distance away from the glass or walls. Side staging should be positioned at least 10 cm (4 in) away from the side to allow air circulation. This is particularly important to prevent the cold air stream due to convection flowing down from the roof and across the staging.

The total wattage required for a glass to ground house can be estimated as follows:

> Total surface area of glass in m^2 x 9 x temperature rise required in $^\circ C$ = total wattage required..........2
> Note that the figure 9 equals the coefficient of heat transfer for the glass ($\approx 5.7\ W/m^2/C^\circ$) multiplied by a factor to allow for fortuitous loss of heat from uncontrollable ventilation, exposure of the site, local weather conditions and other factors.
> For simplicity the glazing bars and frame are regarded as equivalent to glass. If the house is on a very exposed site such as a roof, then the estimated figure should be increased by about one quarter.
> (Total surface area of glass in ft^2 x 1.6 x temperature rise required in $^\circ F$ = loading in Btu/hr which if multiplied by 0.29 gives loading in watts.)

Example:

Glasshouse 10 x 5 m (30 x 15 ft) height to eaves 2 m, height to ridge 3 m.

Area of glass	Sides	$40\ m^2$
	Ends	$25\ m^2$
	Roof	$55\ m^2$
	Total	$120\ m^2$

Temperature rise $13.5^\circ C$ ($25^\circ F$) to maintain a minimum internal temperature of $7^\circ C$ ($45^\circ F$). It is assumed for these calculations that the temperature outside is approximately $-7^\circ C$ ($20^\circ F$).

120 x 9 x 13.5 = 14.6 kW 2

It is essential that the temperature should be controlled by a thermostat. Rod-type thermostats are relatively cheap but much more accurate control is achieved using an aspirated thermostat. This consists of a box in which air is drawn over the thermostat by a small fan. Models are available by which differential day/night temperatures can be obtained. In a small house the aspirated thermostat will help to increase air movement.

Thermostats should be mounted in such a position that they are not unduly affected by the heat from the sun and give a reasonable control of temperature so that the difference in various parts of the house is not too great.

Lighting

Light is required in the house to allow work to be carried out in late winter afternoons and also to supplement the natural light to the plants.

Lighting to allow work need not be of a high intensity unless this involves close examination of material. A maximum intensity value of 300 lux (about 27 fc) measured at the working surface will be satisfactory.

Information on suitable light intensities for plant growth is given in the section on lamp frames, page 24. If automatic timing is installed then a cycle of 12 hours light and 12 hours darkness will be generally satisfactory unless investigations into day length requirements are being carried out.

During the summer months it will usually be necessary to reduce the radiation entering the house by some form of shading. This can be achieved by the use of special glass in the sunny side of the house, an expensive method, or the use of plastic blinds inside the house or slatted wooden rollers usually fixed outside. The application of a shading wash to the outside of the glass is probably the best method for school use. Proprietary washes such as 'Summer Cloud' are available or mixtures of whiting or slaked lime in water can be applied. It is usually only necessary to coat the sunny side of the roof. The wash must be removed during the winter. Since some proprietary brands are difficult to remove the whiting/water mixture is often preferable as most of it will have been washed off by the end of the summer and removal of the rest is relatively easy.

Air-movement

Air circulation is essential for healthy plant growth since it promotes a uniform temperature distribution and reduces the incidence of fungal infection. If tubular heaters are used then, especially during the winter when ventilation may be reduced, small fans may be necessary to create an adequate movement.

Ventilation is necessary to achieve satisfactory temperature control, it also creates air-movement. Hinged ventilators in the roof of the house should be such that the area opened by their maximum movement (excluding the side area) is at least equal to one sixth of the total floor area of the house, preferably as much as one quarter. This is a standard rarely achieved, especially in small domestic greenhouses in which temperature control may become a serious problem in the summer. Indeed, in small houses ventilators in the side may be essential to give a sufficient 'chimney' effect to achieve satisfactory ventilation. As a general rule it is always better to have more than adequate ventilation!

In small houses hinged roof ventilators can be controlled by mechanical devices depending on the expansion and contraction of a material with a high coefficient of expansion operating through a lever system. These are relatively cheap and can be adjusted to give various settings. For the size of house recommended, ventilation controlled by a suitably positioned rod thermostat or connected to the aspirated thermostat box is preferable. Fans should have automatic shutters and be placed in the end of the house away from the entrance and near the ridge. The ventilator thermostat should be set to operate at a difference of 5°C (9°F) above the operating temperature for the heating system during the spring and winter but kept at about 13° C (55°F) in the summer. The capacity performance of the fan or fans should be such that when working a maximum of 10–12 air changes per hour is achieved. To allow for their servicing or in the event of breakdown it would be wise not to rely entirely on this mechanical ventilation but to have in addition adequate ventilators in the roof which can be used in an emergency.

Humidity

This is best controlled by having an adequate reservoir of water available within the house in the soil and the composts in pots and trays. Additional water can be sprayed in and this will often be necessary in summer, especially in those houses which are on roofs or relatively exposed. For most plants, with the exception of cacti and other xerophytes, a relative humidity of between 60–70% is suitable during the growing season. This may be measured using a hygrometer.

EQUIPMENT

In a fully equipped greenhouse the following items will be needed:

1 staging–this should be of slatted wood, preferably cedar or other wood resistant to decay. Metal supports are preferable, they must be corrosion resistant. Staging may not be necessary all round the house (see page 33). The height must be suitable for student use, consequently it is often necessary to cut down the supports which are supplied.
2 capillary trays or beds–these consist of large plastic, metal or glass-fibre trays up to 10 cm (4 in) deep in which the water level in sand, gravel, similar artificial material or synthetic capillary matting which fills the tray is kept constant. This may be achieved in a variety of ways, the water units being independent of or connected to the water mains (see Fig. 7). Pots used on capillary benches should not have stones or broken clay pot 'crocks' in their base (see page 39). Capillary trays filled with sand or similar material are very heavy and need strong supports. Capillary matting is advantageous since it is so light.
3 overhead mist watering electrically controlled or a trickle or drip irrigation system in which water is fed to beds, straw bales or moist capillary beds through perforated black capillary tubing.
4 mist propagating area–this consists of a bed similar to a capillary bed with low voltage soil warming cables embedded halfway down in a layer of sand 10–15 cm (4–6 in) deep at a loading of about 162 W/m^2 (15 W/ft^2). In this area a film of water is maintained over the foliage of leafy cuttings by spraying a fine mist of water at intervals. Cuttings are either rooted directly into the sand on the bed or into medium boxes standing on the sand. Control is either by an electric timeclock which is set to give a short burst of water at intervals irrespective of the weather conditions or by an 'electronic leaf'. This is a sensing device which is positioned in the bed amongst the cuttings to receive the mist spray and which switches this on when its surface reaches a particular level of dryness. Mist spray nozzles are usually placed about a metre (1 yd) apart and about 37.5 cm (15 in) above the bed or according to the makers instructions (see Fig. 17).
5 automatic fertiliser distributors–these are placed

Figure 17 Small mist propagation area showing spray nozzle and sensor device. The base is heated and the area surrounded by polythene film supported on wooden uprights.

in the water supply line of a drip feed or trickle water system and inject fertiliser solution at predetermined rates.

6 automatic pest control—electrical devices are available in which pesticide chemicals are heated and evaporate into the air thus keeping a suitable level of the chemical in the atmosphere and depositing it on the plants (see page 37).
7 maximum and minimum thermometer.
8 hygrometer.

MAINTENANCE

The following principles should be adhered to, to maintain the best standards of plant husbandry.

1 Purchase, or collect when necessary (see page 37), only those plants which, as far as is known, are free from pests and disease.
2 Ensure that all plants are growing in suitable fresh compost or medium in pots or containers of a size adequate for their roots or other environmental needs, and that they receive a sufficient quantity of water and nutrients.
3 Control the environment within the house so that it is kept within the recommended limits.
4 Inspect the plant stock at regular intervals, replacing where necessary, and incinerating those plants which are diseased or very heavily infested with surface pests and damaged.
5 Ensure a high standard of hygiene by removing dead parts of plants and incinerating these or effectively composting them at a high temperature and by keeping the interior of the house clean.
6 Use approved pesticides at regular intervals to control plant infection and disease.[46,49,57]
7 Regularly check all electrical fittings and equipment for safe function.
8 Maintain the greenhouse and its fittings, services and equipment in good condition. Clear the greenhouse once a year, thoroughly wash with disinfectant/detergent solution and fumigate with sulphur dioxide. Note that this is only possible if the entire plant stock can be temporarily removed.

Chapter 4 Cultural directions

Handling plants is a good deal less hazardous than handling animals. In fact the major dangers from managing plants arise from the necessary maintenance techniques.[14,34] It is activities such as spraying with pesticides that really create problems if not done sensibly.

Being aware of safety is a valuable experience for children. The aim of any safety measure is to eliminate hazards or, if this is not possible, to reduce them to an acceptable minimum. Children and other people have to be trained to recognise and to avoid or deal with hazards. They must also be protected by the use of safe equipment and adequate protective devices. Protective clothing must not be overlooked.

The most important physical health hazard is that of infection. The handling of dirty equipment which is likely to cut or abrade the skin leads to a direct risk of infection, as does the contact of dirty hands with the mouth. Cuts and wounds can become secondarily infected from dirty equipment or from composts and must always be protected by a dressing. In some areas of the country there is a risk of tetanus if open wounds are contaminated with soil. Accidents involving soil and open wounds should be referred to hospital or medical authorities. An anti-tetanus injection may be necessary.

Some people have a hypersensitive reaction to certain chemical substances called allergens. These allergic reactions may be recognised by a reddening and possible swelling of the skin after contact or by a increased flow of secretions in the nose and eyes and possible respiratory distress, such as coughing or difficulty in breathing, if the allergen is in suspension in the air. A familiar instance of the latter is hay fever. Several common plants including *Chrysanthemum,* and *Primula* cause contact allergic reactions in the skin. Individuals known to be sensitive to these plants would be wise to avoid contact with them. Dormant bulbs purchased for planting are often treated with fungicide, insecticide, acaricide or other powders to which some people are allergic.

Irritation of the skin and mucous membrane, as distinct from an allergic reaction, may occur in response to a variety of chemicals including many used as pesticides and herbicides and, for example, from the small spines of some cacti which penetrate the skin when the plant is handled and then break off.

Poisons are a hazard usually only when ingested or inhaled. Poisons are found in the seeds and fruits of many common plants including holly, laburnum, privet and yew[35,36,37,38,39] in addition to those commonly used in the teaching of biology, for example, the castor-oil seed. Many seeds, particularly those bought from agricultural suppliers, may be dressed with poisons used as pesticides or bird repellants. Many pesticides and herbicides are, or contain, dangerous poisons. Furthermore some of these chemicals, if not poisonous, may induce cancers (carcinogens). Their use must be strictly controlled (see page 38).

When buying or collecting plants or seeds they should be, as far as is known, free from pests and disease to restrict the spread of infection to other plants. Plants which are usually propagated vegetatively, for example, the potato, most soft fruits and other fruits should be from virus-free stock.

PESTICIDES

Broadly speaking there are four groups of pests which affect plants—weeds, that is unwanted plants in the wrong place, micro-organisms including fungi, viruses and bacteria, nematode worms and arthropod animals including insects and mites.[46,47,49]

All these can be controlled, if not eliminated, by correct and hygienic maintenance techniques. The importance of hygiene cannot be overstressed and the educational value of using hygienic procedures is considerable. Plant stock must be regularly inspected for signs of pests or disease and preventive measures must be taken if infected plants are found. These may involve the isolation of the plant, removal and incineration of affected parts, repotting in sterilised compost or medium, or chemical control.

There are chemical preparations which may be used to control or kill these groups of pests. Many are dangerous to people and so must be used with great care, especially in schools. The Ministry of Agriculture Fisheries and Food operates the Agricultural Chemicals Approval Scheme (ACAS) whereby manufacturers' recommendations for the use of their pesticide products may receive official approval. (See Appendix 2 Section 1 for address for information.) Such products may be recognised by the presence of a label consisting of a large A surmounted by a crown. This scheme is entirely voluntary and only a few products in use are not currently approved under this scheme. It would be wise for teachers to check that all such chemicals they use are approved. All chemical pesticides must be used according to the manufacturers' instructions. All containers and cans in which solutions have been prepared must be thoroughly washed out after use since even very small traces of some of these chemicals, in particular the herbicides, can affect plant growth.[40]

To spray solutions of pesticides, pressure sprayers may be used which are manually pumped up before use or, since many are now packed in aerosol cans, these may be most convenient. Pressure sprayers present particular hazards. In children's hands there is a temptation to squirt the solution in all directions. They should be used only under constant supervision. Aerosol cans must be disposed of with care when empty.

In the greenhouse suitable pesticide chemicals can be continuously released into the atmosphere from electrical devices which heat small pellets of the chemical. If children are working in such an environment for considerable lengths of time then such devices may be thought to be inappropriate. Control of insects, mites and fungi may be conveniently achieved by the use of 'smoke' pellets or canisters. These are ignited releasing the chemical as a smoke which passes to all parts of the house. They are very effective in action but should not be used by pupils. After having the house closed for a suitable period to allow the particles to settle after 'smoking' it must then be thoroughly ventilated before anyone enters.

FERTILISERS

Fertilisers are substances containing the chemicals required for plant growth and are broadly of two types, organic and inorganic.

Organic or natural fertilisers are derived from living organisms and include manure, dried blood, bone meal, spent mushroom compost and so on. They are usually applied in a dry form and in relatively large quantity. Sometimes solutions or suspensions in water may be made from them and watered or sprayed on. The mineral elements that they contain vary enormously. In general whilst they may have a high nitrogen content they are often deficient in others so as to be unbalanced (see pp. 40, 41, 65). These nutrients are usually slowly released over a long time and organic fertilisers may therefore be described as slow acting. Apart from their fertiliser effect those bulky forms such as manure alter the physical condition of the soil in that the water retention, drainage and aeration are all improved.

Inorganic or artificial fertilisers are usually 'man-made' from constituent chemicals. Examples include ammonium salts, potash and urea, the last often being applied as a foliage spray. Artificial fertilisers are usually sold in powder, crystalline or pellet form, sometimes as a solution. Application is generally easy, especially if they are pelleted, but must be done with care to ensure that they are evenly distributed and not spread dry on foliage. Unlike organic fertilisers, they have a quick action and can obviously be formulated to give a balanced compound fertiliser.

The relative proportion of nitrogen (N), phosphorus (P) and potassium (K) is often described as the plant nutrient ratio. For example, if a nutrient or fertiliser mixture has a plant nutrient ratio of 1:1:1, it means that it contains equal parts of nitrogen (calculated as N), phosphates (as P_2O_5) and potash (as K_2O) taken in that order. Similarly, a nutrient ratio of 3:2:1 means that there are three parts of nitrogen and two parts of phosphate to one part of potash and so on.

Nitrogen induces vigorous growth. Phosphorus and potassium are necessary for the physiological activities of the plant promoting balanced growth.[63] Since adequate quantities of phosphorus sufficient for the life of the plant can be incorporated in the growing medium or compost its incorporation in a nutrient solution may not be necessary. A 3:0:1 plant nutrient ratio will promote rapid growth and may be given to young plants. A 1:0:1 ratio is suitable for older plants, finishing with a 1:1:1 ratio to ensure satisfactory final growth and maturity.

In addition to these major elements fertilisers should contain other elements in very small quantities. These are called 'trace elements' and include, for example, boron, iron and magnesium. Trace elements are usually present in organic fertilisers but may be absent from many artificial ones.

GROWING SEED PLANTS IN CONTAINERS

One of the most important factors in the successful growing of plants in pots is the compost or medium in which they are grown. It should have adequate water retention, good aeration and drainage, a sufficient and balanced supply of plant nutrients available and be free from diseases, pests and living weed seeds. Composts such as the John Innes series or soil-less peat based media are recommended. (See Appendix 2 section 2.3 for suppliers.)

The conditions in the soil and whether the plant nutrient is organic or inorganic, affect the availability of the nutrients for the plants. Under certain soil conditions such as high acidity or alkalinity, despite their presence, plants may be unable to absorb some elements, particularly the trace elements, which are thus 'unavailable'.

The John Innes seed and potting composts (see Appendix 1 page 65) are soil-based. Since the soil used will vary and the quality of this is critical, absolute uniformity cannot be obtained. Most plants may be started in John Innes Potting Compost No. 1 and then transferred to Compost No. 2 in larger pots. Vigorously growing plants may be finally potted into Compost No. 3.[50]

Soil-less peat based media are mainly derived from the original formulae of the University of California[58] and are very uniform; furthermore a single formulation may be used throughout the growing life, from seed to the mature plant. In most cases sufficient nutrients are incorporated which makes further addition unnecessary for the first month or so of plant growth. Such media are light in weight and allow easy penetration of roots. However it is easy to over-water these media since the colour change on watering is not as great as with soil. It is usually better to under-water rather than over-water and to test moisture levels by observing the colour. As a general rule the darker this is the higher the water content, furthermore soil based composts become powdery when dry. All composts and growing media must be stored in a dry covered place so that they may remain sterile.

Plastic or earthenware pots or shallow pans are suitable containers. It is essential that these should have adequate drainage holes and that they are not blocked. A layer of broken pot or small pebbles ('crocks') in the base of the pot may be used to ensure free drainage. These should not be put in if the pot is to be stood on a moist bench or tray since contact between the sand on this surface and the compost or growing medium is essential. If plastic pots are used the frequency of watering may be reduced, but the soil temperature will be higher than that in earthenware pots under similar conditions. To facilitate watering the top of the compost should be 1 cm below the rim of the containers. Shallow pans are recommended for creeping plants such as clovers. Plastic pots should be light brown and not brightly coloured since bright pigments in the plastic are often harmful to plant growth.

The soil water should be adequate to maintain the growth of the plant without any check; insufficient or excess is to be avoided. The compost or medium is watered as soon as the plant is seen to be limp. With some plants with soft foliage which easily becomes infected by fungi and which should not therefore be excessively wetted, it is better to water by standing the pot in a shallow pan filled with water. For convenience, pots in greenhouses may be watered on capillary benches (see page 35) or on a moisture proof growth area filled with sand (see page 21). Excessive transpiration is to be avoided; a suggested humidity range for most plants is between 60–80% relative. Spraying the plants may be beneficial but should not be done when the plants are in full sunlight. For most plants air and soil temperatures should be between 10–20°C (50–68°F) during the day and between 5–15°C (40–60°F) at night.

Adequate day length and light of suitable intensity and mixture of colours to maintain normal growth and flowering of the plant should be present. Plants needing 'short day' lengths of 10–13 hours to flower include *Chrysanthemum* and *Poinsettia (Euphorbia pulcherrima)*, 'long day' plants needing 14–18 hours include *Dahlia* and a majority of annuals. A light intensity of 800–1000 foot candles (8600–10 700 lux) is, in general, adequate for most plants though higher intensities may be advantageous (see page 24). Under the conditions of the teaching laboratory it is more than likely that inadequate light will be available; features indicating this include a pale colour and an abnormal elongation between the points of origin of the leaves.

The regular use of nutrient solutions when watering will assist normal growth and they may be added automatically into the bench watering system (see page 35). Nutrients may also be incorporated in a solution which is sprayed on the foliage but such application must be made relatively frequently. There is a considerable range available commercially either in liquid or powder form. It is essential to follow the instructions provided carefully to avoid possible damage to the plants. These nutrient solutions must contain a suitably balanced proportion of available nitrogen (N), phosphorus (P) and potassium (K) and trace elements (see page 38).

When the plant has grown so that its roots almost fill the pot it should be 'potted on' to a larger pot. Restricting the root growth by keeping the plant in small pots may however accelerate flowering but usually reduces its length of life. For the final potting it will be rarely necessary to use a pot bigger than 12.5 cm (5 in) maximum internal diameter, for most plants a 10 cm (4 in) pot will be adequate. If the plant is to remain in one pot for any length of time then some of the compost or medium from the top of the pot should be removed and replaced with fresh material at least once a year.

Some air circulation is essential, but plants should not be in direct draughts as these cause abnormal transpiration rates which can seriously affect growth.

Competition with other living organisms must be strictly limited. Regular spraying with a suitable insecticide and fungicide solution is recommended. It is essential to follow the instructions provided with pesticide chemicals to avoid possible damage to the plants, and it is advisable to avoid the use of persistent organo-chlorine, organo-phosphorus and other similar compounds. A dilute solution of soap or detergent is often perfectly effective in removing surface pests such as aphids and also fungal spores; it also cleans the surface which is usually necessary with plants kept indoors.

Dead leaves and dead unwanted flowers should be removed and burnt. The tips of the shoots should be cut off or 'pinched-out' when convenient. This has the effect of stimulating the growth of lateral shoots from the angle between the leaves and main stem thus maintaining the plant in a compact shape. It also delays flowering and, with experience, may be used to time onset of flowering if this is necessary for class use or for show purposes. The pieces removed may be used as cuttings.

During the summer months and holidays most pot grown plants may be placed out of doors. If this is done then some protection against damage by birds may be needed and semi-hardy species must be protected if frost is expected.

With those plants normally dormant in the winter, watering and nutrient supply should be gradually reduced in the autumn thus inducing a quiescent period when little growth takes place.

Further information about individual plants may be found in references 59, 60, 61, 62, 64, 69, 71, 72, 73, 78, 79, 81, 82, 83, 86 and 87.

WHY POTTED SEED PLANTS FAIL TO GROW

To diagnose the reasons for failure in correct growth, the indicative signs which represent departures from normal growth and development should be considered first.

Bear in mind, however, that it is rarely worth while spending too much time in an attempt to save unhealthy plants. They may, moreover, be a hazard to the health of other plants, if infected with disease or pests.

Indicative signs

Plant-organ	*Sign*
Stem.	Weak and relatively thin. Elongated intervals or internodes between leaf stalks. Abnormal colours. Growth deformity.
Leaves.	Whole leaf an abnormal colour, possibly yellowish. An abnormal colour on parts of the leaf, possibly in spots. Leaf tip, or other parts, withered. A wilted appearance. Leaf fall at an abnormal time of season. Growth deformity.
Flowers.	Failure to form flowers. Failure of flower buds to open. Premature fall of the flower buds, the open flowers or their parts. An abnormal colour on parts of the flower, possibly in spots. Growth deformity.

If some of these indicative signs are evident, then each of the possible causes should be considered in the sequence given.

Possible causes of failure

Watering.	Is the quantity adequate or excessive? Has the compost been allowed to dry out completely and the plant to wilt? Has the pot been standing in water? Has the plant been over-watered during the winter or a 'rest' period? Is the water acidity/alkalinity (pH) satisfactory? Is there too much dissolved calcium or chlorine? Is the water clean? Is the water temperature satisfactory?
Light.	Is the intensity satisfactory? Has the daylength been adequate? Has the plant been exposed to full sunlight, possibly with insufficient water present? Is the mixture of colours in the light satisfactory?
Atmosphere.	Is the humidity satisfactory? Has the humidity been constant? Is the ventilation satisfactory? Has the plant been in a draught? Are there noxious gases or vapours present? Has a smoke fumigant been used incorrectly? Has condensed water been dripping on to the plant?
Temperature.	Is it too high or too low? Is there much variation in temperature, particularly the air temperature, possibly caused by draughts etc.? Is the plant too near a heater? Is there a considerable difference between the temperature of the root and that of the shoot? Has the plant been placed in full sunlight and the pot been sun baked? Has the plant been subjected to frost, possibly as a result of being placed near a window? Has the temperature been inappropriate to the seasonal requirements of the plant?
Compost or growing medium.	Is the mixture correct for the species or cultivar of the plant? Were the components of good quality and adequately mixed? Does it contain traces of hormones or herbicides? Is the drainage free and satisfactory? Is the acidity/alkalinity (pH) satisfactory? Is there an adequate supply of available nutrients? Is the temperature satisfactory? Is the population of other living organisms in the compost too large? Are there large animals e.g. earthworms, present, causing disturbance of the roots? Is the compost or medium old?
Nutrients.	Is the balance between the major nutrients nitrogen, phosphorus and potassium correct? Have trace elements been absent or unavailable? Has an adequate supply been continuously available? Too much lime? Were nutrient solutions added in excess when the plant was newly established? Has a nutrient solution been added when the compost was very dry? Has a nutrient solution been added when the plant should have been in a 'resting' period? Has a dry powdered nutrient been in contact with the plant for some time? Has a foliar 'feed' been incorrectly applied?
Pests.	Is the plant 'clean' and free from external pests? Is the plant free from internal parasites? Has the shoot been damaged by birds?
Disease.	Is the plant diseased? (Infections by fungi and viruses are the commonest; bacterial infection is also possible.)
Spraying.	Has the plant been incorrectly sprayed with water for example when in full sunlight? Has the plant been incorrectly sprayed with a solution of pesticide, either at a concentration higher than recommended or which is not recommended for the species? Has a herbicide solution been in contact with the plant or compost?
Mechanical damage.	Have the roots been damaged for example during transplantation?

GROWING SEED PLANTS IN THE OPEN

The same basic requirements are needed to grow plants in the open as to grow them in pots. However the methods used to achieve these are quite different, especially in the treatment of the soil. The purpose of cultivating the soil is to achieve adequate water retention with sufficient drainage and aeration to have a balanced supply of plant nutrients always available, and to reduce or eliminate competition by weeds and other pests.

It is rarely necessary to dig the soil to a depth greater than a spade's depth, say 30–45 cm (12–16 inches). When starting the cultivation of new ground a thin layer of the immediate surface of the soil containing growing weeds and seeds is first removed and may be composted. The area is then dug over, the first trench of soil being removed and placed to one side. The soil beneath this trench, and each successive one is broken up with a fork and organic compost incorporated. The top soil from the next trench is broken up with the spade and placed as the surface layer of the previous trench and so on. Finally the soil that was removed from the first trench is placed in the last. Dry artificial fertiliser powder or pellets may be incorporated in the surface soil during these operations.

Only those deep tap roots such as those of dandelion or the creeping roots of bindweed or ground ivy, need to be removed before the roots of other weeds are chopped up with the spade. This technique not only improves the physical and chemical conditions in the soil but also controls weed growth, at least initially. Subsequent digging to this pattern need only be done at irregular intervals when the site is clear.

Weed control can be achieved mechanically by hoeing, by heat using a flame gun or chemically using suitable herbicides. Clearly the latter two methods must be done with extreme care! There are various chemicals now available which kill the plant when absorbed through the leaves yet are not absorbed through the stems, particularly if woody, and are inactivated on contact with the soil. Some of these are selective herbicides in that they kill weeds without seriously affecting the crop (see page 37). These require very careful application and all the cans and containers in which solutions have been must, when finished with, be thoroughly washed.

Further details of the specialist techniques to ensure the optimum conditions for individual plants are given in references 69, 70, 79, 83, 84, 85, 87.

PROPAGATION

Seed

Seeds should be sown in John Innes Seed Compost or a soil-less medium. It is convenient to use standard plastic horticultural seed boxes which can easily be covered with a flat sheet of glass or plastic or with a transparent domed plastic cover.

The seed box should be filled with compost or medium to about 1 cm below the rim. The surface is then gently firmed down and the material watered by standing the box in water in a shallow container. After it is thoroughly moistened it must be allowed to drain.

The seeds should be sown evenly over the surface and then covered with a layer of compost or medium. Alternatively a nail board can be used to make depressions in the surface into which the seeds are placed (see Figs. 18 and 19).

The seeds should be covered or planted to a depth approximately equal to their maximum dimension. They must not be sown too deep, a maximum of 1 cm is suggested for most plants. A suitable cover should be placed on the seed-box to restrict water loss and the box kept in the dark until germination takes place. A suitable temperature range for germination is 20–22°C (68–72°F) and bottom heat may be necessary (see page 21). When the shoots appear the covers on the seed-boxes must be removed to allow air circulation. The use of a fungicide spray at this stage will be beneficial in restricting or preventing fungal infections. As soon as the seed-leaves or the first leaves of the shoot have opened out the seedlings must be transferred or 'pricked out' into another container such as a deeper seed-box or pot (see Fig. 20). It is advantageous to transfer the seedlings within a day or two of germination and important not to damage their roots during the process.

If plants are grown on in whalehide or peat pots, or soil blocks, they may be planted out complete with this container which will eventually disintegrate in the soil.

If the plant is to be grown in a pot it may be necessary to 'pot on' the seedling when it is established into larger pots (see Fig. 21).

Figure 18 Broadcast sowing of seeds. Seeds are evenly distributed over the surface of the compost and may then be covered with a layer of sand. This appears lighter in the photograph.

Figure 19 Spaced sowing of seeds. (a) Depressions are made in the soil using a simply constructed nail board wich fits into the seed tray.

(b) Individual seeds are then placed in the depressions and the soil surface smoothed over.

Figure 20 Pricking out. (a) Seedlings are removed from the first container in which they are grown.

(b) Individual seedlings separated with minimum root damage and inserted into depressions in the soil.

(c) The soil is then packed around each seedling to give stability.

Figure 21 Potting on. Individual plants are placed into plant pots partially filled with soil and further soil added around them. If the plants are to be planted out it is advantageous if they are potted into whalehide or peat 'pots'. The entire 'pot' and plant can then be planted later thus minimising root damage.

Stem cuttings

Impatiens, Pelargonium, Tradescantia, Zebrina and many other herbaceous plants are easily propagated from stem cuttings. The most convenient time for this will be during the spring and early summer. If the plant has a woody stem then cuttings are usually taken later in the season. Short pieces of shoot are removed by cutting just below the point of origin of a leaf (a node) with a sharp knife. Some of the leaves at the base of the 'cutting' should then be removed to reduce water losses (see Fig. 22). The cuttings are inserted into holes in compost mixed with coarse sand or soil-less medium. This may be placed in a seed-box, pot or polythene bag. Reducing the water loss and raising the humidity by covering the container will facilitate rooting. Seed-boxes or pots with cuttings may be conveniently enclosed in polythene bags. Alternatively cuttings may be rooted in large trays under a mist-propagator (see pp. 35, 36).

Cuttings from many plants for example *Impatiens, Tradescantia* and *Zebrina*, will form roots when they are standing in water. Indeed this is often the simplest way of propagation. The water should be changed, the container cleaned and the moist end of the cutting gently washed say once a week.

Gentle bottom heating of the compost or medium accelerates the production of roots. Rooting of cuttings inserted into compost or medium may also be speeded up by the use of a hormone rooting

Figure 22 Stem cuttings. (a) A short piece of stem is detached and the lower leaves removed so that water loss is reduced.

(b) The prepared cutting is then inserted in compost.

(c) Adventitious roots form at the cut surface.

chemical. Most cuttings from herbaceous plants will have rooted within two to five weeks at a temperature of 20–25°C (68–77°F). Excessively high air temperatures should be avoided, and frequent inspection made to check on possible fungal or insect attack. For further information see references 66, 68 and 80.

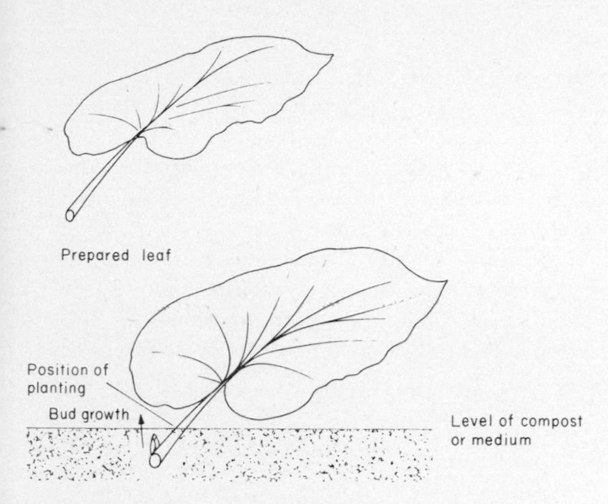

Figure 23 (a) Leaf stalk cuttings of Saintpaulia.

Figure 23 (b) Leaf cuttings of 'Rex' Begonia.

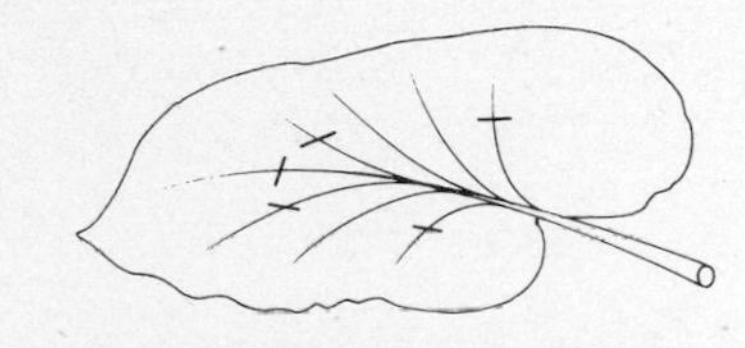

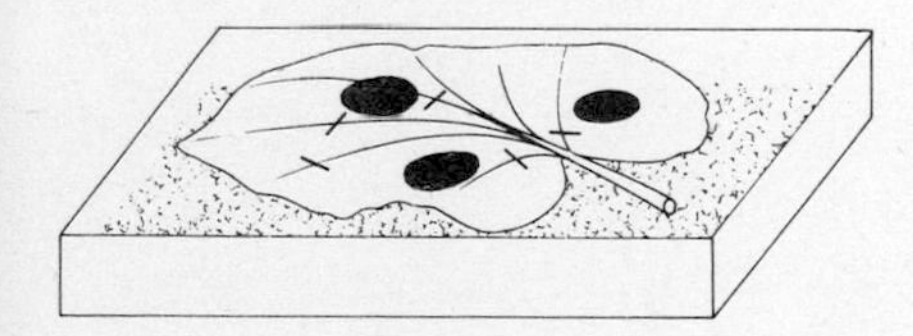

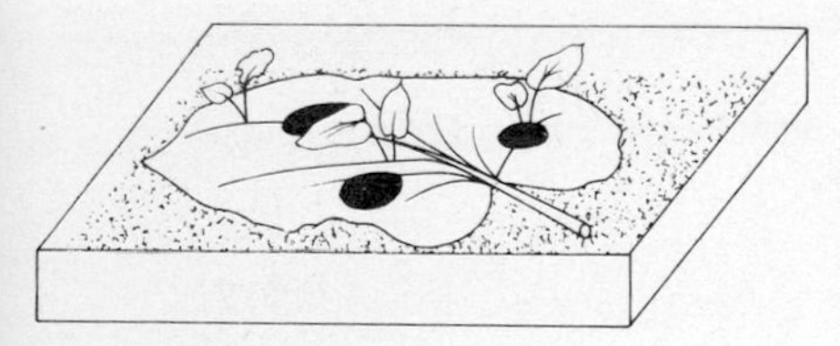

Root cuttings

Cuttings from roots may be made in a similar manner to those from stems and inserted in compost or medium. Root cuttings are usually made in early spring.

Leaf cuttings

With a few plants, for example *Saintpaulia* (South African Violet) detached mature leaves with the leaf stalk attached can be rooted as described using the stalk as if it were a stem cutting[78] (Fig. 23).

Other plants for example the 'Rex' begonias can be propagated from leaf cuttings. In this case a mature leaf is detached and most of the leaf stalk cut off. The underside of the veins are then cut at intervals of 4–5 cm (1.5–2 in) and the leaf placed flat on and in contact with the compost or medium. Roots and new plantlets which form at these cuts are then separated and 'potted on'.

Division of plant or offsets

Large clumps of a plant for example *Chrysanthemum;* offsets, for example *Agave;* or bulb scales, for example *Lilium* can easily be propagated by division (Fig. 24). The best time for this is just before the initiation of growth in the spring. Alternatively it can be done just before the onset of dormancy in winter. The divided plants are then planted out or potted separately.

For further information see references 66 and 80.

Figure 24 Offsets. Young plantlets rooting whilst still attached to the parent plant. The plantlets may be detached and planted separately.

Figure 25 (a) Air layering.

Figure 25 (b) Stem layering in Dianthus.

Layering

Certain plants for example *Ficus* may be 'air-layered'. A ring of the outer layer of the stem is carefully cut away. Around this cut surface moist sphagnum moss or 'Vermiculite' is packed within a piece of polythene which is tied to the stem above and below the cut. Roots form at the cut surface. In general this is a slow process possibly taking up to six months.

With *Dianthus* and other plants the long stems can be half cut through and then buried in the soil compost or medium, without detachment from the parent plant. Roots again form at the cut surface (see Fig. 25).

For further information see references 66, 68 and 80.

Budding and grafting

In both of these techniques the active growing tissue in the bud or graft (scion) is brought into intimate contact with that of a disease free root-stock plant. Growth of the bud or graft starts, if the tissues are compatible, soon after union occurs using the water and minerals absorbed by the roots of the stock. This root-stock plant has a considerable influence on the development of the bud or graft, for example dwarfing root stocks are commonly used with fruit trees.

These techniques are particularly useful with many woody plants. In these the active growing tissue used occurs as the innermost layer of the soft outer tissue surrounding the harder woody centre of the stem (see Fig. 26).

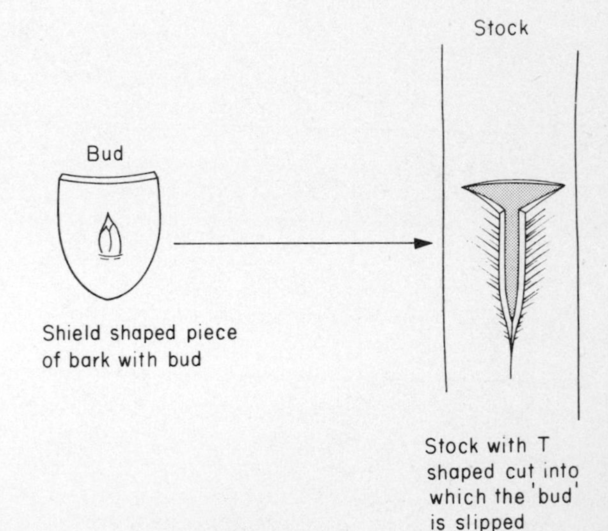

Figure 26 (a) Budding.

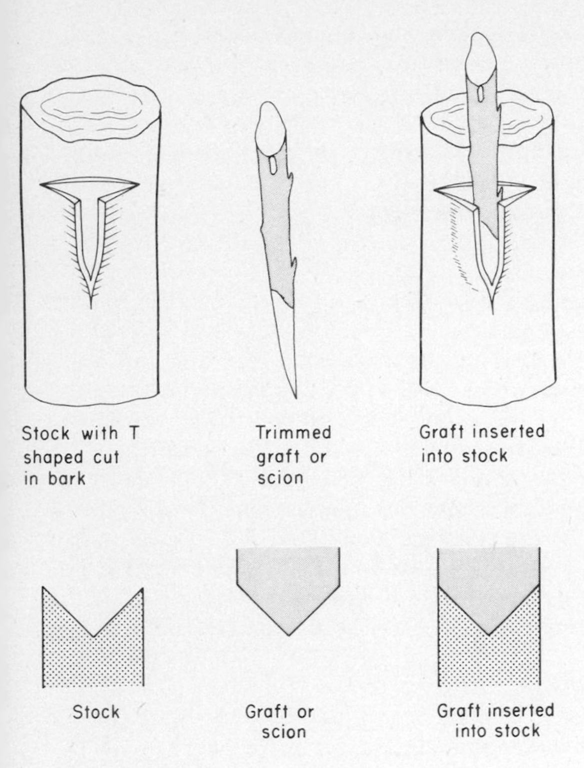

Figure 26 (b) Grafting.

With budding, in for example the rose, a shield-shaped piece of outer tissues with a bud in its centre is taken and inserted into a T shaped cut in the outer tissues of the briar stock. The piece is held in position to the stem by binding, using raffia or special grafting wax. The wax and, to a lesser extent, the raffia also, prevents infection. The binding must be loosened as soon as growth starts.

Grafting can be done in a similar manner. A terminal length of stem of the plant to be grafted is removed. The cut end is obliquely cut so that it can be inserted into a T shaped cut in the outer tissues of the stock. The graft is held in position by binding or using wax. If the diameter of the stem of graft and stock are approximately equal then the cut end of the stock may be cut to an inverted V shape. A corresponding V shaped cut is made in the graft so that it can rest on top of the stock. They can be supported by binding on a length of cane.

Grafting is not commonly used with herbaceous plants. Certain varieties of cacti are grafted and the technique is increasingly used with the tomato.[129]

Further details of these techniques may be found in references 66, 67 and 68.

LIVERWORTS AND MOSSES, FERNS AND AQUARIUM PLANTS

Since the cultural techniques for these are specialised additional information is given for each group.

LIVERWORTS AND MOSSES

As a general rule these plants grow best in a very high humidity and at a low light intensity. They grow very well in terrariums and bottle gardens. A compost of equal parts by volume of loam, leaf mould, peat and coarse sand is recommended. If this is well moistened when the terrarium or bottle garden is set up little further attention is necessary.

They may be conveniently propagated by division. Natural propagation by spores or other reproductive bodies will occur.

FERNS

Mature plants of temperate species will grow out of doors or in an unheated greenhouse.[64,86] The soil should be light and rich containing leaf-mould or peat. For cultivation in pots the compost should be similar to that recommended for liverworts and mosses. It is best to plant or pot the ferns in April and place them in a shady position. If they are kept in a greenhouse it is essential that the humidity should not drop too low; a suggested minimum range is 60–70% relative humidity. If chemicals are used for controlling pests particular care should be taken to ensure that the chemical is suitable for use on ferns and that the concentration and application is correct.

The fern plant is the spore producing phase (sporophyte) in the life-cycle and can be propagated by division or by the spores.

Collection of spores

Fertile fronds of *Pteris aquilinium* (Bracken) or of *Dryopteris felix-mas* (Male Fern) can be collected in July or August when most of the sporangia in which the spores develop will be mature but unopened. The sporangia grow underneath the leaves protected by a covering which will appear glistening and dark brown or black at this stage. Suitable fronds should be collected early in the morning and transported in a polythene bag. Those parts of the leaves which appear ripe should be cut off, placed in a small paper bag and left in a dry atmosphere for about a month to allow the spores to be discharged. These will appear as a brown powder and, with other debris, can be stored in dry tubes. At 2°C (36°F) the spores will remain living for several years.

Before sowing, the spores should be separated from other debris by sieving, first through three or four layers of butter muslin and then one or two layers of coarse lens tissue.

The structure that develops from fern spores is the phase in the life cycle in which sex-cells (gametes) are produced; it is called the gametophyte. When fully developed this is small, flat, green and heart-shaped. The sex-cells are produced in cavities on its under surface. When they are mature the gametophyte must have a film of water covering the under-surface. Into this the male sperm are released and swim to the cavities in which the female egg-cell has developed. After fertilisation a small spore-producing fern is produced which completes the life cycle.

Germination of spores

To have early gametophyte stages available for display the spores should be sprinkled on fluid medium (see Appendix 1 page 66) in Petri dishes and the lid replaced. The development of the fern gametophyte is very sensitive to the wavelength of light falling upon it. Red light inhibits the transition from a filament of cells to the final flattened structure. The dishes should be kept at room temperature in dim light. The spores will germinate in about a week.

To have later gametophyte and early sporophyte stages available for display the spores are best grown on a solid agar medium (see Appendix 1 page 66) in Petri dishes. Spores should be sprinkled on the surface of the medium and the treated plates kept at room temperature in sealed polythene bags or in plastic boxes with some moist cotton wool to ensure that they do not dry out. Low light intensities are again advisable. The heart-shaped gametophytes should develop in about seven or eight weeks after sowing, the sporophytes in about ten.

For general cultivation spores are best sown on moist peat blocks. These should be stood on a flat sheet of glass or plastic and covered with an inverted plastic aquarium or box. Low light intensities are again advisable. At 20°C (68°F) the spores take about ten to twelve weeks to develop into a small fern. During this period it is essential that the peat blocks do not dry out. They should be moistened with a nutrient solution. The young ferns should be transplanted attached to a small piece of peat to avoid damaging the roots.

Further information may be found in references 64 and 86. Details of a sterile technique for germinating spores can be found in reference 55.

AQUARIUM PLANTS

A layer of planting medium in which various species are rooted should be placed in the base of the aquarium tank. This should consist of coarse sand or a mixture of loam, peat and coarse sand. Leaf mould, compost or artificial fertiliser must not be added since nutrient chemicals will dissolve in the water and cause problems with excessive growth of algae. The water should be added to the aquarium tank a few days before the plants are introduced so as to allow noxious gases to be released. It should be poured onto a sheet of paper, placed on the surface of the compost to avoid disturbing the planting medium. Once the plants have been introduced the aquarium should be left once more for a few days to settle before animals are introduced.

Suitable plants include *Alisma* spp., *Elodea* spp., and *Vallisneria* spp.[62]

Chapter 5 Use

Plants are found in most schools, the commonest being the resilient easily grown plants. Cut or dried flowers and foliage are features of temporary decoration. A fortunate few schools will have flower beds or gardens in their grounds—less than 20% of primary and 50% of secondary schools.[8] Many children pass through gardens on their way to and from school or will visit botanic gardens, arboretums or other plant collections. Some will grow plants at home or help with the gardening there.

The local Parks and Gardens Department will usually be only too pleased to arrange visits to their greenhouses or propagating units where the many activities necessary for the proper growth of plants can be observed. Similarly visits may be arranged to local horticultural shows, though these usually occur on Saturdays or during the holiday periods.

To make the most of the various educational opportunities arising from these 'contacts' between pupil and plants, their peculiar use must be carefully related to appropriate aims and objectives which will depend in the main on the pupils' age, ability, interests, attitudes and sometimes on their sex.

GENERAL PRINCIPLES

The aims in any educational use of living organisms must be considered against a background of current educational thought. It is now usually accepted that the child should, to a large extent, work on problems of his or her own choice, that these problems should be taken from the child's own environment, as far as possible, and approached in a practical way. Teachers should be individually responsible for presenting what in their opinion are suitable experiences, to their own pupils which will enrich and satisfy the requirements of the child.

Suggested aims for using living organisms are given in reference 14. Those appropriate to the use of plants are:

1 to provide a source of inherently interesting material which can be used to arouse and encourage an attitude of controlled curiosity and inquiry.
2 to provide the opportunity for personal experience in observing and investigating living organisms, their diversity, their variation, their inter-relationships and life processes.
3 to inspire and encourage creative work in a variety of disciplines.
4 to promote an understanding of some of the concepts of biology and of the processes associated with life.
5 to identify and to examine those factors in the immediate environment of living organisms which affect them and to develop an understanding of the relationships between living organisms and their environment.
6 to promote an understanding of the relationship between man and other living organisms; of his dependence on many living organisms; of the reasons for his exploitation of certain living organisms for food and for other needs; and of his competition with pests and predators.
7 to develop sensitivity to and consideration for the needs of living organisms.
8 to encourage an attitude of concern about the conservation of national environments in which living organisms may thrive, and an awareness of the problems associated with conservation.
9 to develop an aesthetic appreciation of the colour, form and movement associated with living organisms and of the enjoyment of them.
10 to give information about and experience in the necessary skills involved in the techniques of care and management of living organisms.
11 to give opportunities of taking responsibility for the welfare of living organisms.

As a result of a questionnaire enquiry by the Project in both primary and secondary schools, and by questioning teachers at other levels of education, it was possible to devise a grouping for the use of living organisms, under four headings.[5,14]

Focus for Biological Enquiry
Use as a focus for enquiries involving a scientific approach to problems and their solution and as illustrative material for the study of living things.

Centres of Interest and Activity
Use as a source to inspire creative work in a variety of disciplines which are essentially non-scientific and may be described as 'arts and crafts'.
Associated Work in other Subjects
Use as an integral part in an approach to mathematics, to economics, to the physical sciences and to studies leading to historical and geographical investigations.
Remedial and other Beneficial Uses
Use of certain groups of living organisms, especially flowering plants and small mammals, for the general beneficial effect it appears to have with most children. Exceptional children who are deprived in some way, maladjusted, educationally sub-normal or who lack a major sense, for example sight, may be particularly helped in a remedial way by their association with living things, both plants and animals.

Hazards to both mental and physical health may occur in the use of living organisms. The latter are the more likely when plants are used.[54]

Emotional disturbances may arise, especially with the exceptional child who is mentally disturbed or retarded, when plants, which have been cared for by these children die, or are damaged by other children. Such events can be very distressing to the exceptional child, although for the majority of children they would appear to be of little consequence. People on the whole apparently have minimal sympathy for, or identification with plants[5] and their mistreatment or death has little effect, if any, on most of us. This lack of feeling can and does lead to indifference to plant welfare; and where potplants are neglected—particularly in a school situation—it may be that undesirable attitudes to other living organisms will develop. From the child's point of view, does the neglect of potplants reflect and convey an attitude of insensitivity to the needs of other living organisms—possibly other people?

Hazards from radiation can arise from radioactive substances used for example in radioisotope labelled solutions or gases. Except for the use of specified low intensity sources, teachers have to be appropriately qualified and specially trained for a minimum period of time in health physics and in laboratory work with open and closed sources.[14,34]

The uses described in this chapter are grouped under the four headings given above. Further uses, particularly for biological enquiry, are described in the references listed in Appendix 3.6

FOCUS FOR BIOLOGICAL ENQUIRY
Observation of plants gives rise to many questions. In many cases these will be useful starting points for experimental work. The correctness of suggested explanations (or hypotheses) may be established by the critical interpretation of information gained from such work. By subjecting hypotheses to experimental investigation we can seek to develop a 'scientific approach' to solving problems.

In doing this the structure and physiological working of plants can be related both to their function and the way they enable the plant to fit into (or be adapted to) its environment. This includes both physical factors and other living organisms.

It is convenient to recognise that plants can be used, under this heading, in two main ways. Firstly, since they are living organisms, they can be used as illustrative or investigative material for the fundamental characteristics and activities of living things. For example the basic properties of protoplasm, enzyme action, and nutritional requirements. Secondly they can be used to illustrate the unique characteristics of green plants and various groups into which they are divided.

The range of personal investigations and observations that can be made in schools is obviously limited by many factors—finance, availability of facilities, and so on. However, many can be supplemented by the use of second-hand or recorded information whilst still maintaining the enquiry-orientated approach. Such uses involve physiological data, photographs, models, preserved material and mounts and microscope slides.

1. Physiology
1.1 Germination

Seeds to demonstrate the different methods of germination and food storage are sown in peat or vermiculite or sandwiched between the wall of a glass beaker and absorbent paper which is kept moist.

Onion	*Allium*	Monocotyledon, endospermic, epigeal
Mustard	*Brassica*	Monocotyledon, non-

		endospermic, epigeal. This is particularly good for root hairs.
Pea	*Pissum*	Dicotyledon, food stored in cotyledons, hypogeal.
Castor oil	*Ricinus*	Dicotyledon, endospermic, epigeal.
Broad bean	*Vicia*	Dicotyledon, food stored in cotyledons, hypogeal.
Maize	*Zea*	Monocotyledon, endospermic, hypogeal.
Tree seeds	Various species	Tree seeds often need 'frosting' by being left out in the winter.[4, 98, 128]
Pollen tube germination		
Daffodil Polyanthus	*Narcissus* *Primula*	Grow in sucrose solution.[95,111,115]
Chickweed	*Stellaria*	Growth on stigma. Collect in morning and make squash mounts.[115]

1.2 Growth

Duckweed	*Lemna*	Measure increase in surface area of water covered or increase in number of plantlets.[120]
Broad bean	*Vicia*	Demonstrate using plumule growth.
Mung bean	*Phaseolus*	Demonstrate using root growth.[112]
Poinsettia	*Euphorbia*	Leaf colour and response to day length.
Oat	*Avena*	Effect of plant hormones on the growth of the coleoptile.[92,117]

1.3 Basic Properties of Protoplasm

Canadian pond-weed	*Elodea*	Streaming of protoplasm in mesophyll cells.
Snowberry	*Symphoricarpus*	Large cells in fruit show cell structure with prominent nuclei, vacuoles and streaming in protoplasmic strands.
Spiderwort	*Tradescantia*	Streaming of protoplasm in strands across vacuoles in cells of staminal hairs Osmotic relations.[92]
Beetroot	*Beta*	Permeability of the cytoplasm.[92]
Onion	*Allium*	Epidermal cells of leaves in bulb useful to demonstrate cell structure.

1.4 Sensitivity and Tropisms

Onion	*Allium*	Inhibition of germination by light and response of plumule to light.[103]
Venus fly trap	*Dionaea*	Quick movement in response to touch.
Sensitive plant	*Mimosa*	Quick movement in response to touch.
Wood sorrel	*Oxalis*	Photonastic sleep movements.
Wild clover	*Trifolium*	Photonastic sleep movements.
Maize	*Zea*	Photosensitivity of coleoptile and response to hormones.[112]
Cress	*Lepidium*	Response of plumule to light.[112]
Garden pea	*Pisum*	Response of plumule to light.[112]
Sundew	*Drosera*	Response of leaf hairs to touch.
Mung bean	*Phaseolus*	Response of radicle to gravity.[112]

1.5 Transpiration

Willowherb Privet Geranium	*Epilobium* *Fuchsia* *Hydrangea* *Ligustrum* *Pelargonium*	Cut stem of rooted plant attached to manometer to show root pressure. Shoots in manometer for transpiration.[92,116,127]

1.6 Nutritional requirements

Duckweed	*Lemna*	Grow plants in various nutrient solutions in Petri dishes. Count number of

		plantlets at intervals. Illuminate in uniform light under fluorescent tubes about 15 cm above the dishes.[120,125]
	Conocephalum *Marchantia*	Growth of gemmae in nutrient solution.[130]
Wheat Maize	*Triticum* *Zea*	Grow in thoroughly washed sand with the addition of nutrient solutions.[92]
Venus fly trap Sundew Butterwort	*Dionaea* *Drosera* *Pinguicula*	Insectivorous plants.
1.7 Photosynthesis		
	Abutilon	Variegated leaf to demonstrate necessity of chlorophyll for photosynthesis.
Geranium	*Pelargonium*	Production of and testing for starch in leaves.[92,108,116]
	Iris	Production of sugars in leaves.[92,116]
Canadian pond weed	*Elodea*	Release of oxygen.[92,110]
1.8 Respiration		
Pea Castor oil	*Pisum* *Ricinus*	Grow seeds to demonstrate aspects of respiration. Ratio of oxygen intake to carbon dioxide release.[92,108] Energy changes in respiration.
1.9 Enzyme action		
Barley Castor oil Wheat Maize	*Hordeum* *Ricinus* *Triticum* *Zea*	Enzymes may be extracted from germinating seeds by grinding them with water followed by centrifuging to remove cell debris and starches.[101] Cut germinating seeds may be placed on starch agar for the enzyme to diffuse out.
Potato	*Solanum*	Enzyme which catalyses the build-up of starch from glucose phosphate may be extracted as outlined above.[108,116] Oxidase enzymes may be demonstrated using a solution of guaicum resin.[92]

1.10 Food storage		
Beet	*Beta*	Sugar in tap root.
	Dahlia	Insulin crystals in root tuber cells.[92]
Jerusalem artichoke	*Helianthus*	Starch in root tubers.
Pea	*Pisum*	Proteins in cotyledon.
Castor oil	*Ricinus*	Oil in endosperm.[116]
Potato	*Solanum*	Starch in stem tubers.
Broad bean	*Vicia*	Proteins in cotyledons.[92]
Maize	*Zea*	Starch in endosperm, protein in embryo.
2 Morphology		
2.1 Bryophyta		
Liverwort	*Conocephalum* *Marchantia*	Morphology of liverworts.
Moss	*Funaria*	Morphology of moss.
2.2 Pteridophyta* and *Gymnospermae		
Horsetail	*Equisetum*	Homosporous.
Clubmoss	*Selaginella*	Heterosporous.
Bracken	*Pteris*	External morphology.
Malefern	*Dryopteris*	Particularly useful for demonstrating sporangia.
Larch	*Larix*	Deciduous conifer.
Pine	*Pinus*	Evergreen conifer for details of root, stem, leaf arrangement and cones.
Silver fir	*Abies*	
Maidenhair tree	*Ginkgo*	
2.3 Angiosperms, Monocotyledons and Dicotyledons.		
Lily Wheat	*Lilium* *Triticum*	'Typical' monocotyledons. Monocotyledon with

Maize Dragon tree	*Zea* *Dracaena*	secondary thickening.
Wallflower Sunflower Buttercup Broad bean	*Cheiranthus* *Helianthus* *Ranunculus* *Vicia*	'Typical' dicotyledons

2.4 Structure of root, stem, leaf

Cress Buttercup Maize	*Lepidium* *Ranunculus* *Zea*	Fibrous root structure.
Broad bean	*Vicia*	Tap root structure.
Heather Heath Pine	*Calluna* *Erica* *Pinus*	Roots with mycorrhiza.
Holly	*Ilex*	Simple leaf—monocotyledon
Privet	*Ligustrum*	Simple leaf—dicotyledon
Ash	*Fraxinus*	Pinnate leaf
Horse chestnut	*Aesculus*	Palmate leaf; shoot and buds
Mustard Melon Nasturtium Artichoke } Sunflower } Dragon tree Sweet flag	*Brassica* *Cucurbita* *Dahlia* *Tropaeolum* *Helianthus* *Dracaena* *Acorus*	Stem structure

2.5 Special Adaptations

Roots

Celandine	*Ranunculus* *Dahlia*	Root tubers.
Beet Carrot Radish Parsnip	*Beta* *Daucus* *Raphanus* *Pastinacea*	Tap roots.
Virginian creeper Ivy	*Ampelopsis* *Epiphyllum* *Hedera*	Aerial and adventitious roots.

Stems

Virginian creeper Ivy Hop Pea	*Ampelopsis* *Clematis* *Hedera* *Humulus* *Pisum*	Adaptations to climbing and trailing.
Water plantain	*Alisma*	Aquatic stem.
Candle plant Butchers broom	*Kleinia* *Ruscus*	Cladode.
	Crocus	Corm
	Iris	Rhizome
Strawberry	*Fragaria*	Runner
Blackberry	*Rubus*	Stolon
Bamboo Mint	*Arundinaria* *Mentha*	Suckers.
Potato	*Solanum*	Tubers.
Cactus	*Opuntia*	Xerophytic modification.

Leaves

Privet	*Ligustrum*	Mesophytic leaf.
Onion Autumn crocus	*Allium* *Colchicum*	Bulb
Lily Hyacinth	*Lilium* *Hyacinthus*	Mesophytic monocotyledon leaf.
Daffodil	*Narcissus*	
Duckweed Water lily	*Lemna* *Nymphaea* *Nuphar*	Hydrophytic leaf.
Aloe Heather	*Agave* and *Aloe* *Crassula* *Erica* *Euphorbia* *Sedum*	Xerophytic modification.
	Vallisneria	Aquatic leaf.
Venus fly trap Sundew Butterwort	*Dionaea* *Drosera* *Pinguicula*	Modification of leaves to catch insects.
Sweet pea Grape vine	*Lathyrus* *Vitis*	Modification for climbing.
	Bryophyllum	Plantlets on leaf edges.

Sensitive plant	*Mimosa*	'sensitive' leaves

Flowers

Cauliflower	*Brassica*	

2.6 Flower Structure

Onion Lily Wandering jew	*Allium* *Lilium* *Tradescantia*	Simple actinomorphic monocotyledon
Buttercup Wallflower	*Ranunculus* *Cheiranthus*	Simple actinomorphic dicotyledon
Orchids	*Bletia* *Pleione*	Zygomorphic monocotyledon
Milfoil Monkshood Snapdragon Columbine Sweet pea	*Achillea* *Aconitum* *Antirrhinum* *Aquilegia* *Chrysanthemum* *Lathyrus*	Zygomorphic dicotyledon
Broad bean	*Vicia*	
Oat Barley Wheat	*Avena* *Hordeum* *Triticum*	Grass
Pot marigold	*Calendula* *Chrysanthemum*	Composite flower
Hazel	*Corylus*	Catkins for wind pollination

2.7 Dispersal–Seed and Fruit; Spore; Spore dispersal

Whilst almost all the plants in the list are suitable the following will be found most useful: *Antirrhinum; Aquilegia; Avena; Clematis; Digitalis; Epilobium; Fragaria; Fraximus; Impatiens; Lathryus; Lunaria; Pelargonium; Quercus* and *Rosa.* To demonstrate spore dispersal the following can be used: *Conocephalum; Dryopteris; Equisetum; Funaria; Marchantia; Pinus; Pteris; Selaginella.*

3. Anatomy & Microscopic Structure

3.1 Root

Angiospermae			
Onion	*Allium*	Monocotyledon	Good fibrous roots from base of bulb.
Bluebell Hyacinth	*Endymion* *Hyacinthus*	Monocotyledon Monocotyledon	
Bulbous buttercup	*Ranunculus*	Dicotyledon	
Mustard	*Brassica*		Growing seedlings on the inside of clay plant pot standing in shallow water for root tip squash and root hairs.[114]
Broad bean	*Vicia*		Root tip, cap and side shoots

3.2 Stem

Pteridophyta		
Bracken	*Pteris*	Rhizome & stem
Gymnospermae		
Silver fir Larch Pine	*Abies* *Larix* *Pinus*	Secondary thickening in two or three year old seedling stems or branches of older trees
Angiospermae		
Monocotyledons		
Bamboo	*Arundinaria*	
Reed	*Arundo*	Special pith cells.
Dragon tree	*Dracaena*	Secondary thickening.
Bluebell	*Endymion*	
Mares tail	*Hippurus*	

Hyacinth	*Hyacinthus*	
Maize	*Zea*	
Dicotyledons		
Sunflower Nasturtium Buttercup	*Helianthus* *Tropaeolum* *Ranunculus*	herbaceous stems
Elder	*Sambucus*	Heavy stem with good pith and lenticels
Beech Rose Lime	*Fagus* *Rosa* *Tilia*	Woody stems

3.3 Special conducting tissues

Silver fir Pine	*Abies* *Pinus*	Bordered pits in xylem
Marrow	*Cucurbita*	Large sieve tubes
Grape vine	*Vitis*	Callus on sieve plates
Water plantain Reed Mares tail Reed mace	*Alisma* *Arundo* *Hippurus* *Typha* *Clematis*	Air spaces in pith

3.4 Leaves

Celery	*Apium*	Leaf petiole.
 Privet Pine	*Iris* *Ligustrum* *Pinus*	Good leaves to section.
 Box Holly	*Crassula* *Buxus* *Ilex*	Leaves with waxy cuticles from which good leaf peels may be made using nail varnish.[92,116]
Onion Daffodil	*Allium* *Narcissus*	Epidermis strips off easily for mounting.
Water lily	*Nuphar* *Nymphaea*	Stomata on top surface and air spaces.
Heather	*Calluna*	

3.5 Spore production and reproductive bodies (see also 2.7).

Horsetail	*Equistetum*	Best collected in the wild about July. See 2.2
Clubmoss	*Selaginella*	See 2.2
Bracken Male fern	*Pteris* *Dryopteris*	Sporangia on underside of leaves. Spore germination on damp soil or peat, fluid medium or agar.[97,111]
Maidenhair tree Silver fir Larch Pine	*Ginkgo* *Abies* *Larix* *Pinus*	Male and female cones at various stages of development
	Marsilea	Motile gametes released when sporocarp is put in water.

3.6 Flower structure

Amongst the most suitable examples are:-

Lily Paeony	*Lilium* *Paeonia*	Good buds for sectioning.
Shepherd's Purse	*Capsella*	Anther structure, ovule and ovary development.

3.7 Squashes to show mitosis and meiosis

Mitosis		
Onion Bluebell Star of Bethlehem Fritillary Broad bean Chickweed	*Allium* *Endymion* *Ornithogalum* *Fritillaria* *Vicia* *Stellaria*	Root tips.[92,98,111] Pollen grain development on the style.
Meiosis		
Onion Lily Paeony Wandering jew	*Allium* *Lilium* *Paeonia* *Tradescantia* (= *Rhoeo*)	Buds.[92,98]

4. Life cycle and perennation

4.1 *Annual, biennial, perennial, monocarpic and polycarpic*

Pea Broad bean	*Pisum* *Vicia*	Annual life cycle.
Beetroot Onion	*Beta* *Allium*	Biennial. Biennial and monocarpic.
Bamboo Agave	*Arundinaria* *Agave*	Perennial monocarpic. These may take up to 50 years to flower. Most other plants which live for more than one year are perennial and polycarpic. Once mature they flower year after year.
	Dahlia	Perennial, polycarpic.

4.2 *Pollination and devices to prevent self pollination*

Primrose	*Primula*	Polyanthus, pin-eyed and thrum-eyed types.
Radish	*Raphanus*	Protogynous. Stigma receptive in bud, stamens ripen after flower has opened.
Willowherb	*Epilobium*	Protandrous. Stamens ripen and shed before stigma surfaces are exposed.
Hazel	*Corylus*	Separate male and female catkins. Monoecious.
Cucumber Marrow	*Cucurbita*	Separate male and female flowers. Monoecious.
Willow	*Salix*	Separate male and female plants. Dioecious.
	Arum	Adaptations to ensure cross-pollination by insects.
Pistol plant	*Pilea*	Explosive buds which discharge pollen.

5 Methods of propagation

5.1 *Vegetative means*

Jerusalem artichoke	*Helianthus*	Root tubers
Lesser celandine	*Ranunculus*	Root tubers
Yarrow	*Achillea*	Suckers
	Bryophyllum (=*Kalanchoe*)	Plantlets on leaf edges
Spider plant	*Chlorophytum*	Offsets
	Crocus	Corms
Canadian pond weed	*Elodea*	Winter buds
Strawberry	*Fragaria*	Runners
Eel grass	*Valisneria*	Runners
	Iris	Rhizome
Bracken	*Pteris*	Rhizome
Mint	*Mentha*	Suckers
Blackberry	*Rubus*	Stolon
House leek	*Sempervivum*	Offsets
Potato	*Solanum*	Stem tubers
Onion	*Allium*	Bulb
Duckweed	*Lemna*	Basal buds
Hyacinth	*Hyacinthus*	Bulb
Daffodil	*Narcissus*	Bulb
Ground orchid	*Pleione*	Pseudobulb

See also 2.5 Stems and Leaves.

5.2 *Artificial means*

Stem cuttings

Geranium	*Pelargonium*	and many other plants.
Fig	*Ficus* *Tradescantia*	

Leaves		
Lily	*Lilium*	Scale leaves to produce new bulbs.
African violet	*Crassula* *Saintpaulia* *Sedum*	Petiole/leaf cutting.
Grafting		
Apple	*Pyrus*	Scion of named variety on stock of crab apple.
Tomato	*Lycopersicum*	Scion on virus free root stock.[129]
Budding		
Rose	*Rosa*	Buds on wild briar stock.
		Budding is a common nursery practice for many named ornamental species of flowering shrubs and trees, as each bud of the scion can produce a new plant when budded on a seedling stock or a vegetatively produced stock of a plant of the same species or genus. See page 46 and references 67, 68 and 80
Division		
	Dahlia	Separation of root tubers.
		An inspection of the plant list will suggest many other examples.

6. Systematic collections

6.1 Garden beds with a variety of plants from particular families for example the buttercup family, *Ranunculaceae.*

These are frequently maintained in the larger botanical gardens (Kew, Wisley, Oxford, Cambridge, Edinburgh) and cover a wide range of families. A smaller number of beds can be planted in school gardens or collections may be grown in pots indoors or in the greenhouse.

The genera within the family Ranunculaceae, for example *Aconitum, Aquilegia, Clematis, Delphinium, Paeonia* and *Ranunculus* show a range of habit and floral structure and form a particularly useful systematic collection.

6.2 Variation

Suitable pot grown collections of clearly recognised species of one genus are useful to illustrate the variation in features by which they may be identified. Species of the gencra *Fuchsia,* and *Primula* are particularly suitable.

Collections of cultivated varieties or cultivars may be used to illustrate the effect of man's selection. Cabbage, red cabbage, savoy, sprout, kohl rabi, kale, turnip are particularly suitable as examples of vegetables selected from species within the genus *Brassica.* See reference 98.

Most cultivated plants have had a number of different varieties of the same species selected as 'fixed' cultivars. These often bear little resemblance to the wild species from which they were developed. In addition to the brassicas, good examples include the many named varieties of *Chrysanthemum, Dahlia* and *Narcissus.*

Fixed varieties which are all propagated vegetatively from the original seedling and which usually cannot be grown from seed are called 'clones'. Examples include the many similar clonal stocks of fruits for example the Cox's Orange Pippin apple and *Tradescantia* or *Rhœo.*

6.3 Special hybrids

Seeds of the first filial generation (F_1 seeds) obtained by crossing selected parent plants of the same species are of special interest commercially.[98] They show hybrid vigour and the parents are selected so that the hybrid will show special features eg:—

Eurocross tomato	*Lycopersicum*	An early maturing fruit immune to mildew and greenback.
Brussels sprouts	*Brassica*	F_1 hybrids such as Peer Gynt produced for deep freezing.
Antirrhinum	*Antirrhinum*	F_1 Coronette.

Other hybrids arise from the crossing of plants of different genera eg.

Hawthorn-medlar	*Crataegomespillus*	An intergeneric graft hybrid between the hawthorn (*Crataegus*) and the medlar (*Mespillus*).
Fig-leaved ivy	*Fatshedera*	An intergeneric hybrid between the fig-leaf palm (*Fatsia*) and the ivy (*Hedera*).

6.4 Polyploid

Many new varieties of flowers and vegetables have been produced by inducing polyploidy by treatment with colchicine e.g.:—

Pacific hybrid primulas	*Primula*
Tetrasnaps	*Antirrhinum*
Polyploid radish	*Raphanus*
Beetroot	*Beta*

Other examples of polyploid plants may be found in the companion volume 'Organisms for Genetics'.[99]

CENTRES OF INTEREST AND ACTIVITY

Plants and flowers attract attention and interest. As such they may be very effectively used as the inspiration for a variety of creative work in the arts and crafts.

The first observations that young children make usually lead to comment and discussion, then finally to a desire to communicate both the observations and their opinions. It is in this process of observation that the human senses of sight, touch, smell and hearing are mainly used, taste but rarely. Interpersonal communication involving the use and stimulation of these senses can therefore be explored, using the plant as a point of origin.

Fig. 27 gives a variety of examples of such creative activity essentially involving observation and the communication of impressions and perceptions.

Such activity is particularly of value with younger children and the slower learners in that it can allow them to satisfy the need to express their own ideas in external composition easily and provides practice in manipulative skill. All young children can make

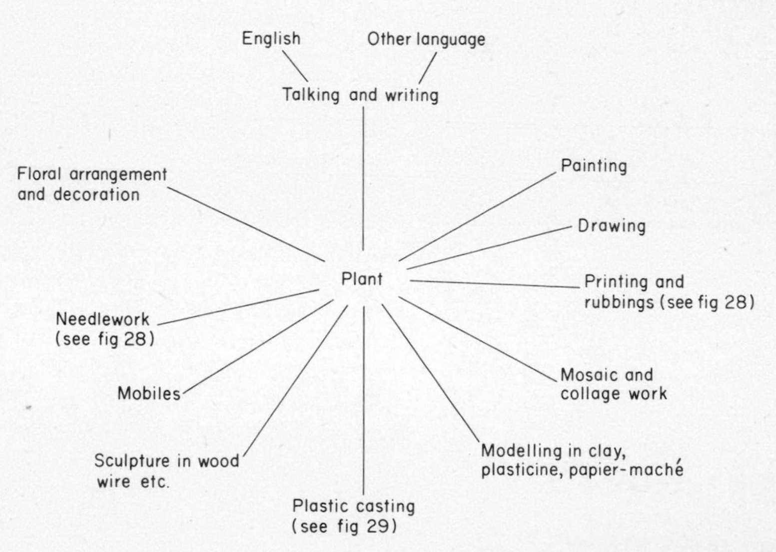

Figure 27 Centre of interest use. Creative activity involving communication.

simple drawings, paintings or models very quickly, enabling rapid expression of thoughts and concepts. Imagination may be allowed a free run, giving rise to fantastic creations which may release inner tensions and be very beneficial. Rubbings, prints or casts can be made from leaves. Leaves and flowers can be dried and preserved. Somewhat older children can write about plants in English or other languages. In the secondary or senior comprehensive school these activities may become more formal, elaborate and complex. Mobiles can be made, needlework done and the possibilities for floral arrangements and decoration are almost endless (see Fig. 28).

Equipment needed for the maintenance of plants may well lead to creative activity in the craft department. Suitable cold frames, stands and other pieces of apparatus may be made from wood and metal. Further information may be found in reference 118.

7. Plants for centres of interest and activity—Arts
Almost any plant may be used under this heading. However certain are particularly useful in view of their appearance or other features.

7.1 Leaf Prints

Maple	*Acer*	Impression prints may be
Painted leaf plant	*Coleus*	made in a manner similar to brass rubbing, using
Hawthorn	*Crataegus*	coloured crayons or 'heel-ball'. Bark rubbings may

*Figure 28 Centre of interest use. (a) Cow parsley plant (*Anthriscus sylvestris*) inspires (b) and (c).*

(b) Silk screen print.

(c) Embroidery.

be made in a similar way. Prints may also be made by first coating the under-surface of the leaf with carbon from carbon paper or with boot polish and then transferring an image from this surface by pressing the leaf to paper. Silhouettes of the leaves may be made by flicking ink or paint, or spraying paint using an aerosol can, around the leaves which are pinned closely to sheets of paper. For further information see reference 131.

7.2 Plaster casts

Ivy leaves Hardfruits and twigs Bark	*Hedera* e.g. *Cornus; Crataegus*	The plant material is half embedded in a thin layer of neutral coloured plasticine, then removed, forming a mould. Walls of plasticine about 1 cm deep are placed round to complete the mould. Fine plaster (e.g. dental plaster) or cement paste is then poured into this and left to set. The cast may be painted and finally varnished as a protection.[131] (Fig. 29)

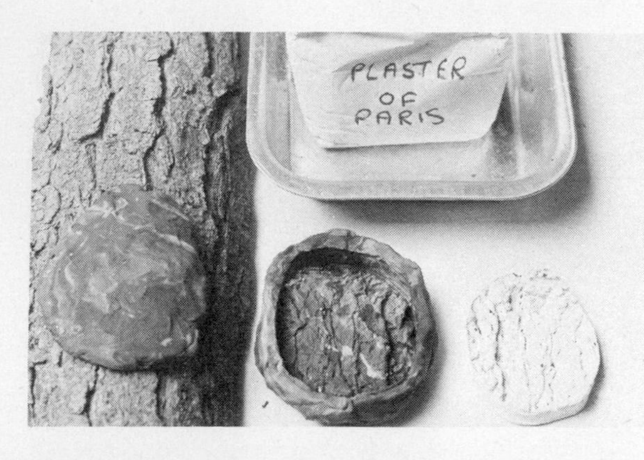

Figure 29 Plaster casts from tree bark. An impression of the bark is made using clay or plasticine and a cast taken from this. The cast can then be coloured if desired. The same principle is used for all plaster casting.

7.3 Plants to dry as specimens or for decoration

Everlasting or straw flowers	*Anaphalis* *Gomphrena* *Helichrysum* *Helipterum* *Rhodanthe* *Statice*	All these flowers or seed heads can be dried by hanging in small bunches upside down in a dry, airy place. Direct sunshine must be avoided, since this can bleach the colour. Everlasting flowers are cut just before they are fully open and the dried flower heads and fruits collected.
Bamboo Oat Sea holly Barley Hop Lily	*Arundinaria* *Avena* *Clematis* *Delphinium* *Eryngium* *Hordeum* *Hydrangea* *Humulus* *Lilium* (seed pods)	Foliage gathered just before autumn, such as beech (*Fagus*), red hazel (*Corylus*) and others, is best preserved using glycerine and then allowed to dry. Sprays of foliage are stood in deep jars containing a 33% glycerine: water mixture for two or three weeks before being removed and allowed to dry.
Honesty Poppy Chinese lantern Bracken Wheat Reed mace Maize	*Lunaria* *Papaver* (seed pods) *Physalis* *Pteris* *Triticum* *Typha* *Zea*	A variety of flowers may be dried to create lifelike effects using silicagel or a mixture of equal parts of borax, fine limestone and sand. The flowers are placed in suitable containers which can be sealed and the material carefully poured in to surround them. The containers are then sealed and left for two to three weeks. The material is then dried in an oven before use. Leaves with prominent strong veins such as laurel and ivy may be 'skeletonised' by immersing them in water and allowing the soft tissues to rot away. Alternatively they can be gently heated in a solution of washing soda and lime. In either case they must be carefully washed after removal and then pressed and dried between sheets of absorbent paper.

7.4 *Plants with decorative foliage*

	Abutilon
Flowering maple	*Acer*
	Begonia cultivars
Spider plant	*Chlorophytum*
Painted leaf plant	*Coleus*
Male fern	*Dryopteris*
India rubber plant	*Ficus*
	Gynura
Pick-a-back-plant	*Tolmiea*
Ivy	*Hedera*
Geranium	*Pelargonium*

Especially the varieties Happy Thought and Caroline Schmidt.

Aluminium plant	*Pilea*
Bracken	*Pteris*
Mother-in-law's tongue	*Sanseveria*
Wandering jew	*Tradescantia*

8. Plants for centres of interest and activity— 'Craft' uses

8.1 *Coppice*

Hazel	*Corylus*	For the production of bean poles and long rods for basket making. Sweet chestnut coppice was used in the past to produce wood for charcoal burning (not listed) and split chestnut fencing. Coppice is a method used to produce many young growths from a tree by cutting off the tree at ground level and allowing a number of shoots to grow. They are harvested when they are the correct thickness.
Beech	*Fagus*	For furniture making.

8.2 *Osiers and Willows*

Willows	*Salix*	These are pruned back hard to produce long whippy growths harvested for basket making. In marshy areas trees are headed back to a trunk 2–3 metres above the ground to form a pollarded tree. Willow canes are harvested from the head of the pollard. Willows grow particularly well on marshy ground or alongside water and schools with this sort of environment can produce their own craft material for basket making.
Bamboo	*Arundinaria*	Split cane.
Reed	*Arundo*	Thatching.
Dogwood	*Cornus*	Coloured wood for basketry.

8.3 *Timber*

Not a practical proposition for growing at school. Established woodland near school can often be used for extraction of timber and many landowners and local authorities welcome the co-operation of schools in maintaining areas of woodland. The Forestry Commission is very helpful in providing opportunities for schools to join in activities in woodland. Schools can also play a valuable part in conservation of woodland, tree planting and setting up nurseries for growing trees from seeds for planting.

Sycamore	*Acer*	Wood for rural crafts; growing winged fruits for planting.
Horse chestnut	*Aesculus*	Trees for planting grown from 'conkers'.
Beech	*Fagus*	Good hardwood for craft and tool making; seedlings from beech mast for planting and hedging.
Ash	*Fraxinus*	Wood for craft; seedlings from ash keys.
Holly	*Ilex*	
Oak	*Quercus*	Felled timber for craft use; growing acorns for planting.
Willow	*Salix*	Soft wood for carving; cricket bat wood.
Lime	*Tilia*	Hardwood for carving.
Larch Spruce	*Larix* *Picea*	Softwoods from conifers.
Pine	*Pinus*	Growing conifer seedlings in a Dunneman bed for planting.

ASSOCIATED WORK IN OTHER SUBJECTS

In many subjects where work of a personal creative nature as already outlined is not involved, use of plants can lead into a variety of enquiry orientated activities, where the teacher can decide how 'open' or 'closed' they should be.

The pupils are directed along particular lines of enquiry which may be essentially information collecting or problem solving or a mixture of both. It will often be impossible to distinguish between the 'non-scientific' and the 'scientific'.

Fig. 30 indicates a number of possible lines of such enquiry.

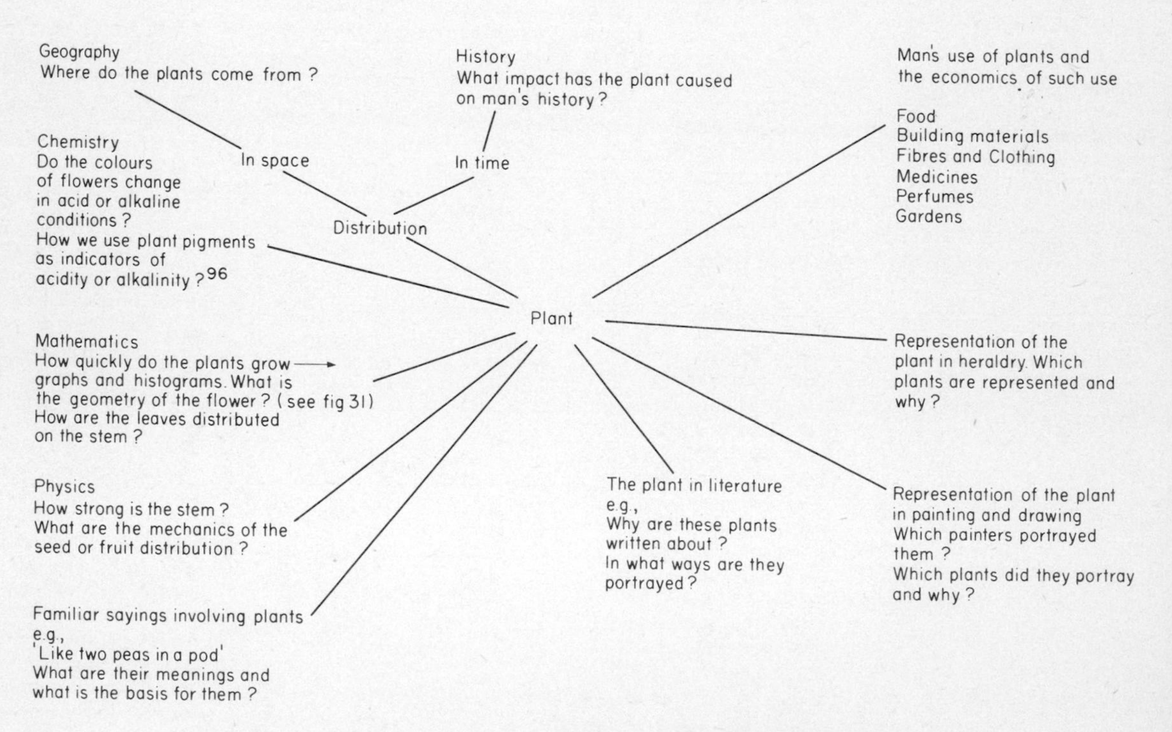

Figure 30 Associated work in other subjects. Possible lines of enquiry.

Figure 31 Associated work. Leaf patterns in Rhododendron. *Is the arrangement in a regular pattern? If so, how can this be expressed mathematically?*

9. Economic Plants

9.1 Market garden crops

Beet Turnip Carrot Parsnip Radish	*Beta* *Brassica* *Daucus* *Pastinacea* *Raphanus*	As these are grown mainly for foodstuffs for man the part of the plant harvested is usually specially adapted for food storage. These are all plants which store food in their tap roots. Apart from the radish these plants are usually biennial but are harvested as tap roots at the end of the first season of growth.
Artichoke Potato	*Helianthus* *Solanum*	Potato and artichoke have food storage organs (tubers) formed from the stem. The stem tubers also serve as organs of perennation and asexual reproduction.
Cucumber	*Cucumis*	
Onion	*Allium*	
Cabbage, kale savoy, sprouts, broccoli, cauliflower, kohlrabi	*Brassica*	All leaf and/or bud storage organs.
Celery	*Apium*	
Strawberry Tomato Pea Blackberry Broad bean Grape vine Maize	*Fragaria* *Lycopersicum* *Pisum* *Rubus* *Vicia* *Vitis* *Zea*	

9.2 Farm plants

Cereals Oat Barley Wheat	 *Avena* *Hordeum* *Triticum*	 Results of selection and varieties grown are discussed in reference 99. Enzymes to digest starch to sugar in brewing are produced by germinating cereal seeds or malt.
Root crops Mangold Swede Turnip	*Brassica*	
Hop Fodder crops Kale Clover Grasses	*Humulus* *Brassica* *Trifolium* Various species	

9.3 Herbs

Mint Marjoram Sage Thyme	*Mentha* *Origanum* *Salvia* *Thymus*	
Yarrow Southernwood	*Achillea* *Artemesia*	Composite herbs
Fennel	Umbellifer herb	Genera not listed in plant list as few have uses other than herbs, with the exception of mint. Herb gardens are frequently grown as special collections.

9.4 Medicinal plants

Foxglove	*Digitalis*	Digitalis

Opium	*Papaver*	Opium
Monkshood	*Aconitum*	Aconitine
Nightshade	*Solanum*	Atropine
Autumn crocus	*Colchicum*	Colchicine
Thorn apple	*Datura*	Largely of historical interest as most drugs are now produced synthetically. A few, including those listed above, are still grown as sources of drugs.

9.5 Tropical plants

Banana Cocoa Coffee Date palm Fig Pineapple Rice	These are grown in most tropical houses in botanic gardens. A collection grown in school forms a useful link with economic geography. Genera not listed.

REMEDIAL AND OTHER BENEFICIAL USES (10)
'The care of growing plants and the keeping and study of animals meets a need that is fundamental to all human beings. The satisfaction of this need consequently enriches the personality and provides an acceptable formative influence on all pupils'.

(Schools Council Working Paper 24. Rural studies in Secondary Schools).

Plants may be used in a therapeutic remedial way with the exceptional child who is mentally or physically handicapped and in addition, with the majority of children, for the general beneficial effect their presence usually has.[5,14] Responsibility for the welfare of plants and the various activities associated with growing them can have a remedial effect with mentally handicapped children. Activities such as seed sowing, potting on and planting out generally have a pacifying, soothing effect. The value of the satisfaction to be gained from seeing plants grow and flourish can be immense. For such children the more robust plants may well be the best including *Calendula, Coleus, Geranium (Pelargonium), Primula,* spider plant *(Chlorophytum),* spiderwort *(Tradescantia),* nasturtium *(Tropaeolum),* and *Tagetes.*

The choice of plants and associated facilities for the child who is physically handicapped is especially critical. With blind children the senses of smell and touch are all important. Plants that have a strong scent, such as *Dianthus* (carnations and pinks), *Hyacinthus* (hyacinth), dwarf varieties of *Lathyrus* (sweet pea), *Mathiola* (stock), *Nicotiana* and *Phlox* cultivars, or have interesting surface contours or textures, such as *Begonia* and many succulents are the most suitable choice. Necessary equipment and composts must be arranged within easy reach.

Children confined to wheelchairs or those with disabilities limiting movement require special facilities rather than particular plants. Attention must be directed towards the elimination of steps and similar obstacles and easy collection of tools. If plants are grown in raised beds built up with a wall of peat blocks to a suitable height and about 1m wide, children in wheelchairs or those who cannot stand for any length of time and must sit can enjoy gardening in much the same way as those more fortunate. Suitable positioning of these beds will prevent congestion and accidents and firm paths between them facilitate movement. Similarly the shelving or staging in greenhouses must be positioned at a suitable height and no more than 0.5m in width or if island staging, 1m wide, to enable easy access. Probably the only plants not suitable are those with a trailing habit. Tools with handles somewhat longer than is usual may be advantageous.

Safety, especially in the use of tools and of electrical equipment, will require special consideration.

Appendix 1 Chemical recipes

The solutions, media and other mixtures which are available ready made from biological suppliers or other sources are indicated by an asterisk.

Bicarbonate/indicator solution

Stock Solution

0.29 thymol blue
0.19 cresol red
20 cm^3 ethanol
0.84g sodium bicarbonate
Analar distilled water
to 1 dm^3

Dissolve the thymol blue and cresol red in the ethanol. It may then be necessary to filter, as dyes from some sources leave a sediment when solutions are prepared. Dissolve the sodium bicarbonate in about 900 cm^3 of the distilled water. Prevent dirt or dust from getting into the water as this will affect the final colour. Add the filtered dye and make up the volume to exactly 1 dm^3. Dilute 1cm^3 of this stock solution with 9cm^3 distilled water for use.

This is a most useful solution to indicate the production of oxygen or carbon dioxide by living things within a closed container. Full details of its use may be found in reference 116. When carbon dioxide is added to the solution the colour becomes yellow. When it is removed the colour becomes purple.

**Composts*

John Innes composts.
These are prepared from the following:

A medium sieved loam partially sterilised to remove harmful organisms. This can be achieved by heating at between 50°C and 100°C for 10 minutes.
B sieved granular or fibrous peat lightly moistened. (Both loam and peat should be passed through a 1 cm sieve.)
C dry coarse sand or grit with particles 3 to 6 mm in diameter.
D superphosphate of lime.
E ground limestone or chalk.
F John Innes base made by mixing well the following parts by weight:
2 parts hoof and horn meal (13 per cent N)
2 parts superphosphate of lime (18 per cent P_2O_5)
1 part sulphate of potash (48 per cent K_2O)

Type	Parts by volume			kg/m³ of the mixture		
	A	B	C	D	E	F
Seed	2	1	1	0.9	0.45	
potting No. 1	7	3	2		0.45	2.4
potting No. 2	7	3	2		0.9	4 7
potting No. 3	7	3	2		1.35	7.0

Preparation.
The loam (A) should be spread out on a clean dry floor with the peat (B) and sand or grit (C) on top. The fertilisers (D, E, F) are sprinkled over the surface to ensure even distribution and the heap turned over three or four times.

Culture solutions for investigations into the effects of nutrient deficiency

Several solutions could be used, the one given being that of Sachs. Only those highest grade chemicals free of contamination must be used.

Basic solution
0.70 g potassium nitrate
0.25 g calcium sulphate hydrated
0.25 g calcium dihydrogen phosphate hydrated
0.25 g magnesium sulphate hydrated
0.08 g sodium chloride
0.005 g iron (III) chloride hydrated
1 dm^3 distilled water

Solutions deficient in certain elements are made by replacing or deleting some of the chemicals listed above.

Minus nitrogen
0.52 g potassium chloride to replace the potassium nitrate

Minus phosphorus

0.16 g	calcium nitrate to replace the calcium phosphate

Minus potassium

0.59 g	sodium nitrate to replace the potassium nitrate

Minus iron

Omit the iron (III) chloride

Minus magnesium

0.17 g	potassium sulphate to replace the magnesium sulphate

Minus sulphur

0.16 g	calcium chloride to replace the calcium sulphate and
0.21 g	magnesium chloride to replace the magnesium sulphate

Minus calcium

0.29 g	potassium sulphate to replace the calcium sulphate and
0.71 g	sodium phosphate to replace the calcium phosphate

Enzyme tests

For oxidases

Freshly prepared solutions of gum guaiacum in ethanol, which are brown in colour, are turned blue in the presence of oxidases.

For starch-splitting enzymes (amylase)

These are likely to be produced in germinating seeds. These should be cut in half and stood on the surface of starch-iodine agar medium in Petri dishes. The area of colourless agar produced below each seed indicates the presence of amylase. Further information may be found in reference 101.

Starch-iodine agar
0.05 g soluble starch
2 g agar
100 cm^2 distilled water
one drop iodine solution

Prepare a suspension of the starch in the water. Add the agar and iodine solution. Autoclave to 121°C (103.5kN/m^2) for fifteen minutes. Allow to cool to 45–59°C before pouring into Petri dishes.

Fern spore germination

Stock solutions

These are best kept at X100 working concentration.

Stock solution 1

60 g	potassium chloride
90 g	magnesium sulphate
36 g	sodium nitrate
6 g	iron (III) citrate
1 dm^3	distilled water

Stock solution 2

100 g	calcium nitrate
1 dm^3	distilled water

Stock solution 3

60 g	potassium dihydrogen phosphate
1 dm^3	distilled water

Fluid medium for use

10 cm^3	stock solution 1
10 cm^3	stock solution 2
10 cm^3	stock solution 3
2970 cm^3	distilled water

Agar-medium

1.5 g	agar
100 cm^3	fluid medium

Add the agar to the medium and agitate. Autoclave to 121°C (103.5kN/m^2) for fifteen minutes. Allow to cool to 45°–50°C before pouring into Petri dishes.

Food test solutions

There are three main groups of food substances; carbohydrates, fats and proteins.

Carbohydrates

Reducing sugars *Benedict's solution (qualitative)

17.3 g	copper (II) sulphate hydrated
100 g	sodium carbonate hydrated
173 g	sodium citrate hydrated
1 dm^3	distilled water

Dissolve the sodium citrate and carbonate in about 800 cm^3 of warm distilled water. Filter and make the filtrate up to 850 cm^3. Dissolve the copper sulphate in about 100 cm^3 of cold distilled water. Pour the citrate-carbonate solution into a large beaker, and add the copper sulphate slowly with constant stirring. Make up the total volume to 1 dm^3.

Test: add reagent to a solution of the substance under test. Heat gently. Production of an orange or red colour indicates the presence of a reducing sugar.

Non-reducing sugars.

Test: boil a solution of the substance under test with 0.07M hydrochloric acid. Allow to cool. Neutralise with sodium bicarbonate and then use test for reducing sugars.
Starch. Iodine solution

1 g	iodine crystals
6 g	potassium iodide

Dissolve the potassium iodide in 200 cm^3 distilled water and then add the iodine crystals. When they are dissolved make up to 1 dm^3 with distilled water.
Test: add reagent to the substance under test. Production of a blue-black colour indicates the presence of starch.

Fats and oils
Test: dissolve the substance under test in isopropanol. Add the solution to water. Production of a milky coloured emulsion indicates the presence of a fat or oil. This test may be made semi-quantitative (see reference 116).

Proteins *Millon's reagent

1 cm^3	mercury
9 cm^3	concentrated nitric acid
10 cm^3	distilled water

Dissolve the mercury in the nitric acid in a small beaker in a fume cupboard. Add the distilled water when the action is complete.

In view of the danger of the components and the hazards of manufacture it is suggested that the prepared solution is bought.
Test: add the reagent to a solution of the substance under test and heat gently. Production of a red colour or precipitate indicates the presence of protein.

Stain solutions

For chromosome preparations
Full details of the appropriate techniques will be found in the companion volume 'Organisms for Genetics'.[99]

*Acetic-carmine

45 cm^3	acetic acid (glacial)
1 g	carmine
55 cm^3	distilled water

Agitate the carmine in the acetic acid, add the water, and bring to the boil. Cool and filter off excess carmine. (This stain may be used with or without a mordant. If a specimen in the stain is teased with iron needles, sufficient iron dissolves from the needles to act as a mordant. A drop or two of 45% acetic acid saturated with iron acetate can be added as the mordant.)

*Acetic-lacmoid
The stain is an indicator and is known as resorcin blue as a dye. Prepare as for acetic orcein.

*Acetic-orcein
The stain orcein deteriorates in dilute acid and so it is best to prepare it from dry stain or keep it as a concentrated solution which must be diluted for use. A greater concentration is required with synthetic orceins.

100 cm^3	acetic acid (glacial)
2.2 g	orcein

Dissolve the orcein in the acid by gently boiling for about 6 hours, using a reflux condenser. Filter and bottle to form a stock solution.
To use mix:

9 cm^3	stock solution
11 cm^3	water

Only dilute sufficient for immediate use.

*Feulgen stain (Schiff's reagent)

200 cm^3	distilled water
1 g	fuchsin basic
30 cm^3	M. hydrochloric acid.
3 g	potassium metabisulphite ($K_2S_2O_5$)

Preparation: Bring distilled water to the boil and add the fuchsin. Shake well and cool to 50°C. Add the M. hydrochloric acid and the potassium metabisulphite. Allow to bleach for 24 hours in a tightly stoppered bottle in the dark. Add 0.5 g decolorising charcoal. Shake thoroughly and filter rapidly through coarse filter paper. Store in a tightly stoppered bottle in a cool, dark place.

Sucrose solutions

For the germination of pollen grains aqueous solutions of sucrose (cane sugar) between 3 and 15% are suitable. For *Narcissus* a 15% solution is recommended, for *Primula* a 5% solution.

If the grains are to be floated in the sucrose solution then 0.02 g of boric acid should be added to each dm^3 of this. If the grains are to be placed on cellophane sheet floating on the solution then 0.5 g of boric acid should be added to each dm^3.

Appendix 2 Sources of information, materials and plants

This list is not completely comprehensive, nor does the inclusion of an address imply that any particular publication, supplier or piece of apparatus is to be preferred to any other. Before buying it is generally advisable to consult the catalogues of more than one supplier.

In order to avoid repetition the address of each organisation or supplier is given in full on the first occasion of mention only. All subsequent references are by name only and with the sub-section number in which the full address may be found. The list is annotated where necessary.

Every care has been taken in compiling this list. However no responsibility can be accepted for any inaccuracies. The list is correct at the time of printing.

It is suggested that teachers, rather than individual pupils, should apply for information.

2.1 INFORMATION

Associations, Societies, Government Departments and Other Sources.

Societies usually produce newsletters and/or annual publications which may contain useful articles on maintenance and give specialist sources of plants.

Agricultural Development and Advisory Service, Great Westminster House, Horseferry Road, London. SW1P 2AE

Agricultural Research Council, Weed Research Organisation,Begbroke Hill, Yarnton, Oxford. OX5 1PF

Alpine Garden Society, Hon. Secretary: E M Upward, Lye End Link, St John's, Woking, Surrey. GU21 1SW

Association of Agriculture, 78 Buckingham Gate, London. SW1E 6PE

Bee Research Association, Hill House, Chalfont St Peter, Gerrards Cross, Buckinghamshire. SL9 0NR

Botanical Society of the British Isles, c/o Department of Botany, British Museum (Natural History), Cromwell Road, London. SW7 5BD

British Tourist Authority, 64 St James's Street, London. SW1A 1NF

Consortium of Local Education Authorities for the Provision of Science, Equipment (CLEAPSE), Brunel University, Kingston Lane, Uxbridge, Middlesex. UB8 3PH. Reports, information sheets and apparatus notes available.

Forestry Commission, 25 Saville Row, London. W1X 2AY

The Electricity Council, 30 Millbank, London. SW1P 4RD

Garden History Society, Membership Secretary: Miss J Lee, 24 Woodlands, North Side, London. SW4 ORJ

Horticultural Education Association, Hon. Gen. Secretary, c/o Pershore College of Horticulture, Pershore, Worcestershire. WR10 3JP

Horticultural Trades Association, Belmont House, 18 Westcote Road, Reading, Berkshire. RG3 2DE

Ministry of Agriculture Fisheries and Food, Tolcarne Drive, Pinner, Middlesex. HA5 2DT

National Federation of Young Farmers' Club, Young Farmers' Centre, National Agricultural Centre, Kenilworth, Warwickshire. CV8 2LG

National Gardens Scheme, 57 Lower Belgrave Street, London. SW1W OLR. Publish yearly 'Gardens of England and Wales Open to the Public'. Addresses of Scottish and Ulster schemes given in this publication.

Pharmaceutical Society of Great Britain, 17 Bloomsbury Square,London. WC1A 2NN

Royal Horticultural Society, Horticultural Hall, Vincent Square, London. SW1P 2PE

Royal Society for the Prevention of Accidents, Cannon House, The Priory, Queensway, Birmingham. B4 6BS

School Natural Science Society, Publications Officer, 44 Claremont Gardens, Upminster, Essex. RM14 1DN

Scottish Schools Science Equipment Research Centre,

(SSERC), 103 Broughton Street, Edinburgh. EH1 3RZ. Scotland. Regular bulletins issued giving information on sources of apparatus and plants. Biological Equipment List published 1970.
Women's Farm and Garden Association, Courtauld House, Byng Place, London. WC1E 7JH

2.2 PUBLICATIONS

Journals, Magazines and Periodicals
Articles on the maintenance or use of plants may be found in many of these publications. The photographs and articles are often very useful illustrative material.

Sources of plants and/or of associated apparatus and equipment are usually given in the advertisement section.

'Agriculture.' Published for the Ministry of Agriculture Fisheries and Food, by Her Majesty's Stationery Office, Editorial Office, Ministry of Agriculture Fisheries and Food (1). Lists chemicals approved by the Agricultural Chemicals Approval Scheme at intervals.
'Amateur Gardening'. 189 High Holborn, London. WC1V 7BA
'Catalogue of the Chelsea Flower Show.' (Yearly) Royal Horticultural Society (1). Contains many addresses of suppliers of plants and horticultural equipment.
'Commercial Grower.' Sovereign Way, Tonbridge, Kent. TN9 1RW
'The Flower Arranger.' The Official Publication of the National Association of Flower Arrangement Societies of Great Britain, 21a Denbigh Street, London. SW1V 2HA
'The Gardener.' The magazine for the north. 39 York Street, Glasgow. G2 8JL
'Garden News.' Park House, 117 Park Road, Peterborough, Northamptonshire. PE1 2TS
'Gardener's Chronicle.' Gillow House, 5 Winsley Street, London. W1A 2HG. Horticultural Trade Journal.
'Grower.' 49 Doughty Street, London. WC1N 2LP. Vegetables; Glasshouses; Fruit and Ornamental plants.
'Journal of the Royal Horticultural Society.' Royal Horticultural Society (1)
'National Federation of Young Farmers' Club. Monthly Newsletter.' (1)
'Popular Gardening.' Tower House, Southampton Street, London. WC2E 9QX
'Practical Gardening.' Mercury House, Waterloo Road, London. SE1 8UL
'Royal Horticultural Society Gardener's Diary and Notebook.' (Yearly). Published by: Charles Letts & Co. Ltd., Diary House, Borough Road, London. SW1 1DW. Lists all Government departments and some associations concerned with horticulture and preservation; botanic gardens; horticultural research stations and specialist horticultural societies.
'Scientific Horticulture.' Journal of the Horticultural Education Association (1)
'Where to Buy: Agricultural and Horticultural Equipment, Supplies and Services.' Where to Buy Ltd., John Adam House, John Adam Street, London. WC2N 6JH. Yearly publication.

2.3 GENERAL BIOLOGICAL SUPPLIERS

These firms supply a wide range of plants suitable for educational use, associated apparatus, equipment and chemicals.

Bioserv Ltd., 38 Station Road, Worthing, Sussex. BN11 1JP
T. Gerrard & Co. Ltd., Gerrard House, Worthing Road, East Preston, Littlehampton, Sussex. BN16 1AS
Griffin Biological Laboratories Ltd., Gerrard House, Worthing Road, East Preston, Littlehampton, Sussex. BN16 1AS
Philip Harris Biological Ltd., Oldmixon, Weston-super-Mare, Somerset. BS24 9BJ

2.4 GLASSHOUSES AND FRAMES

Access, Yelvertoft Road, Crick, Rugby, Warwicks. NN6 7XS. Frames.
Agriframes Ltd., Rollosmatch House, 6 Anerly Station Road, London. SE20 8PT. Plastic cloches. (Plant protection and growing frames).
Alitex Ltd., 30 St Johns Works, Station Road, Alton, Hampshire. GU34 2PZ. Greenhouses and frames.
Alton Glasshouses Ltd., Alton Works, Bewdley, Worcestershire. DY12 1BR. Greenhouses.
Cambridge Glasshouse Co. Ltd., Comberston, Cambridge. CB3 7BY. Glasshouses for commercial growers.
Clear Span Ltd., Wellington Road, Greenfield, Nr. Oldham, Lancashire. OL3 7AG. Commercial and domestic greenhouses.

Crittall-Hope Ltd., Manor Works (Horticultural Department), Braintree, Essex. CM7 6PH. Greenhouses and frames of standard size supplied in kit form.
Edenlite (Greenhouses) Ltd., Hawksworth Industrial Estate, Swindon, Wiltshire. SN2 1EQ. Domestic greenhouses.
Guernsey Glasshouses Ltd., Les Eturs, Guernsey, Channel Islands. Glasshouses.
Hall Robert H, (Kent) Ltd., 5 Church Road, Paddock Wood, Nr. Tonbridge, Kent. TN12 7BR. Greenhouses and frames.
Hancock Aluminium Greenhouses HAG Ltd., Wells Place, Merstham, Nr. Redhill, Surrey. RH1 3AT. Greenhouses and related equipment.
Humex Ltd., 5 High Road, Byfleet, Weybridge, Surrey. KT14 7QF. Automatic propagators; electrically heated garden frames.
Park Lines & Co. Ltd., 717/719 Seven Sisters Road, London. N15 5JV. Greenhouses.
Phillips H E Ltd., King William Street, Coventry, Warwickshire. CV1 5NQ. Greenhouses.
Pluie Frames, Surrey Rose Farm, Guildford Road, Chobham, Surrey. GU24 8EB. Frames and greenhouses.
Pratten, F & Co. Ltd., Charlton Road, Midsomer Norton, Bath. BA3 4AG. Greenhouses.
Q Cloche Ltd., Willowbank Wharf, Ranelagh Gardens, Putney Bridge, London. SW6 3JT. Plastic frames and cloches.
Simpsons of Spalding Ltd., Brantons Bridge, Bourne Road, Spalding, Lincolnshire. PE11 3LP. Commercial glasshouses.
Stewart Plastics Ltd., Purley Way, Croydon, Surrey. CR9 4HS. Plastic cloches.
Walton, E C & Co. Ltd., Old North Road, Sutton-on-Trent, Nr. Newark, Nottinghamshire. NG23 6QN. Greenhouses.
Whitehouse C H, Ltd., Buckhurst Works, Frant, Nr. Tunbridge Wells, Sussex. TN3 9BN. Greenhouses and frames.
Worth Buildings Ltd., Donnington, Telford, Shropshire. TF2 7NF. Greenhouses.

2.5 HORTICULTURAL APPARATUS AND SUPPLIES

Many of the products mentioned will be available from local horticultural sundriesmen and nurseries.

Access (4) Irrigation equipment.
Agriframes Ltd. (4) Plant protection and growing frames.
Associated Sprayers Ltd., Eliot Street, Birmingham, Warwickshire B7 5SS. Horticultural spraying equipment.
Autogrow Ltd., 3-13 Quay Road, Blyth, Northumberland. NE24 2AS. Electrical horticultural equipment. Greenhouse equipment, heaters etc.
Bayliss Precision Components Ltd., Compton, Ashbourne, Derbyshire. DE6 1DA. Automatic greenhouse ventilators.
BEF Products (Essex) Ltd., Old Southend Road, Great Burstead, Billericay, Essex. CM11 2PY. Plastic plant pots and seed trays.
Bondina Industrial Ltd., Greetland, Halifax, Yorkshire. HX4 8NJ. Synthetic capillary matting.
Cameron Irrigation Co. Ltd., Harwood Industrial Estate, Littlehampton, Sussex. BN17 5BR. Glasshouse irrigation equipment; mist propagators; electrical control equipment.
Diplex Ltd., P.O. Box 172, Watford, Hertfordshire. WD1 1BX. Garden and greenhouse thermometers, barometers, frost-predictors, soil moisture meters and allied instruments.
Elt, George H, Ltd., Eltex Works, Bromyard Road, Worcester. WR2 5DN. Greenhouse heaters.
Evenproducts Ltd., Blayneys Lane, Evesham, Worcestershire. WR11 4TS. Irrigation systems.
Fisons Ltd. Agrochemicals Division, Harston, Cambridge. CB2 5HU. 'Levington' compost; 'Plantgrow' compost for indoor plants; pesticides and fertilisers.
House and Garden Automation, 186 High Street, Barnet, Hertfordshire. EN5 5SZ. Greenhouse automation equipment.
Humex Ltd. (4) Automatic greenhouse equipment; heaters; mist propagators; watering systems.
I.C.I. Garden and Household Products Department, Plant Protection Ltd., Woolmead House East, Woolmead Walk, Farnham, Surrey. GU9 7UB. 'Kerimure' soil-less compost; pesticides and herbicides.
Lindsey and Kesteven Fertilisers Ltd., Saxilby, 23 Guildhall Street, Lincoln. LN1 1TR. 'Arthur Bowers' compost; fertilisers.
Manuplastics Ltd., Southdown Works, Kingston Road, Raynes Park, London. SW20 8SD. Plastic plant pots.

Murphy Chemical Ltd., Wheathamstead, St Albans, Hertfordshire. AL4 8QU. Pesticides and herbicides.
Nethergreen Products Ltd., P.O. Box 3, Alderley Edge, Cheshire. SK9 7JJ. Greenhouse and garden watering equipment.
Nelton Ltd., N.E. Wing, Bush House, Aldwych, London. WC2B 4PX. Garden netting; greenhouse shading.
Pan Britannica Industries, Ltd., Britannica House, High Street, Waltham Cross, Hertfordshire. EN8 7DR. 'Baby Bio' soil-less compost; pesticides and herbicides.
Phillips H E Ltd. (4) Greenhouse heaters and soil sterilising equipment.
Plant Protection Ltd., Fernhurst, Haslemere, Surrey. GU27 3JE. Pesticides and herbicides.
Plantpak (Plastics) Ltd., Maldon Road, Mundon, Nr. Maldon, Essex. TM9 6NP. Disposable plastic plant containers.
Q. Cloche Ltd. (4) Plastic plant pots and seed trays.
Sankey, Richard & Son Ltd., The Potteries, Bulwell, Nottingham. NG6 8PE. Plant pots.
Shepherds Aerosols Ltd., Shernfold Park, Frant, Nr. Tunbridge Wells, Kent. TN3 9HH. 'Aerovap' and 'Fumovap' insecticide vaporisers.
Simplex of Cambridge Ltd., Horticultural Division, Sawston, Cambridge. CB2 4LJ. Horticultural instruments; electrical horticultural and greenhouse equipment including heaters, mist propagation equipment, fans and irradiation equipment.
Solo Sprayers Ltd., 4 Brunel Road, Leigh-on-Sea, Essex. SS9 5JN. Agricultural and horticultural hand spraying machines.
Spraygen Sprayers Ltd., 10/12 Carver Street, Birmingham, Warwickshire. B1 3AU. Spraying equipment.
Stewart Plastics Ltd., Purley Way, Croydon, Surrey. CR9 4HS. Plastic plant pots, seed trays and propagators.
Sudbury Technical Products Ltd., 58 Charlton Road, London. SE3 8TT. Soil test kits.
Sutton, C, (Sidcup) Ltd., North Mills, Bridport, Dorset. DT6 3AJ. Fruit cages and netting.
Thermorforce Ltd., Derwent Mill, Cockermouth, Cumberland. CA13 OHS. Automatic greenhouse ventilators.
Tudor Accessories Ltd., Ystrad Mynach, P.O. Box 1, Hengoed, Glamorgan, South Wales. CF8 7XD. Horticultural spraying equipment.
Turner, Stuart Ltd., Henley on Thames, Oxfordshire. RG9 2AD. Electric pumps for garden fountains.
Ward, George (Moxley) Ltd., Baggotts Bridge, Darlaston, Staffordshire. WS10 8QZ. Clay and plastic flower pots; plastic seed boxes.
Woodman, E J & Sons Ltd., High Street, Pinner, Middlesex. HA5 5PN. Horticultural and garden supplies.

2.6 PLANTS

The General Biological Suppliers (3) have a range of plants available; only suppliers of specialised stocks are listed here.

When ordering, adequate prior notice of requirements must be given, particularly for living plants which are usually only available in season.

2.6.1 Seeds and fern spores

The General Biological Suppliers (3) keep a range of seeds suitable for genetic investigations and also irradiated seed.

Seeds from a number of other firms, in addition to some of those listed, are available from local horticultural sundriesmen.

Valuable cultural directions are usually to be found in seed catalogues.

Schemes for the distribution of seeds operated by the major Botanic Gardens, the Royal Horticultural Society and specialist societies are a useful source of some species not readily available commercially.

Burlingham, George, & Sons Ltd., Station Hill, Bury St Edmunds, Suffolk. IP32 6AE. Clover, cereals, herbage and some root crops.
Cannock Fertilisers Ltd., Cannock, Staffordshire. WS11 3LW. Grass seed and fertilisers.
Carters Tested Seeds Ltd., Lower Dee Mills, Llangollen, Denbighshire. LL20 8SD. Also bulbs and garden sundries.
Dobie, Samuel, & Sons Ltd., Upper Dee Mills, Llangollen, Denbighshire. LL20 8SD. *Mimosa pudica*, also fern spores. Bulbs and garden sundries.
Duff, W & Son (Forfar) Ltd., West Craig Nurseries, Forfar, Angus, Scotland. DD8 1XE. Tree seeds (and plants).
Elsoms (Spalding) Ltd., Elsom House, Spalding, Lincolnshire. PE11 1TD. Clover, cereals, herbage and vegetables.

Forestry Commission, Forest Research Station, Alice Holt Lodge, Wrecclesham, Farnham, Surrey. GU10 4LH. Tree seeds.
Howell, Major V F, 'Firethorn', 6 Oxshott Way, Cobham, Surrey. KT11 2RT. Wide range of the more rare seeds.
National Institute of Agricultural Botany, The Official Seed Testing Station for England and Wales, Huntingdon Road, Cambridge. CB3 OLE. Weed seed including *Capsella bursa-pastoris, Cuscuta campestris, Stellaria.*
Pedley, S & Sons, Preston Road Nursery, Newton, Preston, Lancashire. PR4 3RL. *Coleus* seeds.
Practical Plant Genetics, 18 Harsfold Road, Rustington, Sussex. BN16 2QE. Tomato seed for genetics.
Reid, Ben & Co., Ltd., Pinewood Park Nurseries, Countesswells Road, Aberdeen, Scotland. AB9 2QL. Tree seeds (and plants).
Sutton & Sons Ltd., Royal Seed Establishment, London Road, Reading, Berkshire. RG6 1AB.
Thompson & Morgan of Ipswich Ltd., London Road, Ipswich, Suffolk. 1P2 OBA
Unwin, W J, Ltd., Histon, Cambridgeshire. CB4 4LE. Also range of plants.
Vicarage Farm Nurseries, 256 Great West Road, Heston, Hounslow, Middx. TW5 OBN. also at: 141 Vicarage Farm Road, Heston, Hounslow, Middx. TW5 OAA. *Fuchsia* seed.
Wicks, W C Ltd., Lambley, Nottingham. NG4 4QL. *Saintpaulia* seed.

2.6.2 Freshwater (coldwater) plants

Aquatic and semi-aquatic plants.

Many of the firms also supply invertebrate animals and coldwater fish. Other equipment is available as indicated.

Anglo Aquarium Plant Co. Ltd., Wildwoods, Theobalds Park Road, Enfield, Middlesex. EN2 9BP
Bennett's Water Lily and Fish Farm, Chickerell, Weymouth, Dorset. DT3 4AF. Plastic pool liners.
Dewey Waters & Co., Cox's Green, Wrington, Bristol, Gloucestershire. BS18 7QS. Garden pools and accessories.
Earlswood Water Garden, 165 Wood Lane, Earlswood, Warwickshire. B4 5JN. Water garden equipment.
Gerrard & Haig Ltd., Beam Brook, Newdigate, Nr. Dorking, Surrey. RH5 5EF. Garden pools.
Hertfordshire Fisheries, 145 Park Street Lane, Park Street Village, Nr. St Albans, Hertfordshire. AL2 2AX
Highland Water Garden, Rickmansworth, Hertfordshire. WD3 2HB. Fountains; pumps.
Perry's Hardy Plant Farm, Theobalds Park Road, Enfield, Middlesex. EN2 9BG. Ornamental fish.
Shirley Aquatics Ltd., Monkspath, Shirley, Solihull, Warwickshire. B90 4EF.
Stewarts (Ferndown) Nurseries Ltd., God's Blessing Lane, Broomhill, Wimborne, Dorset. BH21 7DF. Pool liners.

2.6.3 Terrestrial Plants

Allwood Bros. (Hassocks) Ltd., Clayton Nurseries, Hassocks, Sussex. BN6 9LX. Cultivars and species of *Dianthus.*
Bees Ltd., Sealand, Chester, Cheshire. CH1 6BA. Herbaceous plants and roses.
Blackmoor Nurseries, Blackmoor, Liss, Hampshire. GU33 6BS. Fruit trees; soft and cane fruits.
Blackmore and Langdon Ltd., Tiverton Hill Nursery, Bath, Somerset. BA2 1NA. *Begonia, Delphinium* and *Phlox.*
Blom, Walter, & Son, Coombelands Nurseries, Leavesden, Watford, Hertfordshire. WD2 7BH. Bulbs and corms (seeds).
Broadleigh Gardens, Barr House, Bishops Hull, Taunton, Somerset. TA4 1AE. Miniature cultivars of *Crocus, Iris, Narcissus* and *Tulipa;* other miniature bulbs.
Clifton Geranium Nurseries, Earnley Gardens Ltd., Cherry Orchard Road, Chichester, Sussex. PO19 2BX. Cultivars of *Geranium* and *Pelargonium.*
Daniels Brothers Ltd., Town Close Nurseries, Daniels Road, Norwich. NOR 55G. Bulbs.
Drake, Jack, Inshriach Alpine Plant Nursery, Aviemore, Invernesshire. PH22 1QS. Alpine plants.
Fisk's Clematis Nursery, Westleton, Nr. Saxmundham, Suffolk. IP17 3AT. Cultivars and species of *Clematis.*
Gregory C & Sons Ltd., The Rose Garden, Stapleford, Nottingham. NG9 7JA. Roses including the miniature cultivars.
Hillier and Sons, Winchester, Hampshire. SO22 5DN. Herbaceous plants, trees and shrubs.
Holly Gate Nursery, Spear Hill, Ashington, Sussex.

RH20 3BA. Cacti and other succulent plants.
Ingwersen, W E Th., Ltd., Birch Farm Nursery, Gravetye, East Grinstead, Sussex. RH19 4LE. Rock-garden plants; dwarf conifers, hardy bulbs and rare shrubs.
Jefferson-Brown, M, Whitbourne, Worcester. WR6 5BR. Cultivars and species of *Narcissus.*
Kelway & Son Ltd., Langport, Somerset. TA10 9SL. Cultivars of *Iris* and *Paeonia.*
Lockyer, C S, 74 Cock Road, Kingswood, Bristol. BS15 2SG. Cultivars and species of *Fuchsia.*
Mackie, A J, Skirmett, Nr. Henley-on-Thames, Oxfordshire. RG9 6TD. Insectivorous plants; hardy orchids and bulbs.
Pedley, S & Sons (2.6.1) Cultivars of *Coleus* and *Hedera.*
Philip Harris Biological Ltd. (3) *Crocus balansae* and a range of species of *Tradescantia.*
Price, Mary E, Fernhurst, Roncarbery, Co. Cork, Ireland. *Asplenium adiantum-nigrum, A. trichomanes, Blechnum spicant, Cterach officinarum, Osmunda regalis, Phyllitis scolopendrium, Polypodium vulgare.*
Robinsons Hardy Plants, Greencourt Nurseries, Crockenhill, Swanley, Kent. BR8 8HD. Rock garden and alpine plants.
Rochford, Thomas & Sons Ltd., Twinford Hall Nurseries, Twinford, Nr. Broxbourne, Hertfordshire. EN10 6BH. House plants.
Rogers, W H (Chandler's Ford) Ltd., Red Lodge Nursery, Chestnut Avenue, Eastleigh, Hampshire. SO5 3HG. Dwarf and slow growing conifers.
Russell, L R Ltd., Richmond Nurseries, Windlesham, Surrey. GU20 6LL. Shrubs; trees; dwarf hardy perennials.
Suffolk Seed Stores, Ltd., Woodbridge, Suffolk. IP12 1DL. Cultivars of *Pyrethrum;* other herbaceous plants.
Sunningdale Nurseries Ltd., The Waterer Group, Windlesham, Surrey. GU20 6LN. Hardy plants and shrubs.
Tokonoma Bonsai, 14 London Road, Shenley, Radlett, Hertfordshire. WD7 9EN. Bonsai trees.
Treasures of Tenbury Ltd., Tenbury Wells, Worcestershire. WR15 3HQ. *Clematis* and alpine plants.
Vicarage Farm Nurseries (2.6.1) Cultivars and species of *Fuchsia.*
Wallace & Barr Ltd., The Nurseries, Marden, Kent. TN12 8BP. Cultivars and species of *Iris* and *Narcissus.* Other bulbous plants.
Waterer, John Sons & Crisp Ltd., The Floral Mile, Twyford, Berkshire. RS10 9SJ. Herbaceous plants and roses.
Wells (Merstham) Ltd., Merstham, Redhill, Surrey. RH1 3AS. Cultivars of *Chrysanthemum.*
Welsh Plant Breeding Station, Plas Gogerddan, Nr. Aberystwyth, Wales. SY23 3EB. *Trifolium repens* with multiple allele leaf markings.
Wicks, W C Ltd. (2.6.1) Cultivars of *Saintpaulia.*
Woolman, H Ltd., Olton Road, Shirley, Solihull, Warwicks. B90 3NQ. Cultivars of *Chrysanthemum* and *Dahlia;* other herbaceous plants.
Worfield Gardens Ltd., Worfield, Nr. Bridgenorth, Salop. WV15 5LN. Bonsai trees.

Appendix 3 Bibliography and references

3.1 General, including conservation and nature trails (1–16).
3.2 Identification, dictionaries and encyclopaedia (17–32).
3.3 Safety (33–39).
3.4 Facilities, control of pests and disease (40–56).
3.5 Maintenance, cultural directions (57–89).
3.6 Use (90–132).

3.1 GENERAL INCLUDING CONSERVATION AND NATURE TRAILS

1 Association of Agriculture (1971) *Bibliography of Sources of Agricultural Material Suitable for Use in Schools.*

2 Bingham C D (1967) *Rural Biology.* Heinemann.

3 Botanical Society of the British Isles (No date) *Code of Conduct for the Conservation of Wild Plants.*

4 British Tourist Authority (1973) *Nature Trails in Britain.*

5 Dennis E (Ed) (1972) *Everymans Nature Reserve. Ideas for Action.* David & Charles.

6 Educational Use of Living Organisms Project (Schools Council) Kelly P J and Wray J D (Eds) (1975) *The Educational Use of Living Organisms. A Source Book.* Hodder and Stoughton Educational.
(1977) *Organisms in Habitats.* Hodder and Stoughton Educational.

7 Forestry Commission (1971) *See Your Forests.* Forestry Commission/HMSO.

8 Kelly P J and Wray J D (1971) The educational uses of living organisms. *Journal of Biological Education* 5 (5) 213–18.

9 Mountain M F (1965) *Trees and Shrubs Valuable to Bees.* Bee Research Association.

10 Postlethwait S N and Enochs N J (1967) Tachyplants–Suited to Instruction and Research. *Plant Science Bulletin* 13 (2) 1–5.

11 Richards A J (1972) The Code of Conduct: A list of Rare Plants. *Watsonia* 9 (1) 67–72 (see also Reference 2).

12 Salisbury Sir E (1962) The Biology of Garden Weeds. *Journal of the Royal Horticultural Society* 87 (8) 338–50; 87 (9) 390–404; 87 (11) 497–508 also printed as a separate publication from the Royal Horticultural Society.

13 Schoolmaster Publishing Co Ltd (1975) *Treasure Chest for Teachers. Services available to Teachers and Schools.* Schoolmaster Publishing Co Ltd.

14 Schools Council (1974) *Recommended Practice for Schools relating to the Use of Living Organisms and Material of Living Origin.* Hodder and Stoughton Educational.

15 Wilson R W (1974) *Useful Addresses for Science Teachers* 2nd Edn. Edward Arnold.

16 Women's Farm and Garden Association (1967) *Guide to Less Common Plants for Use in Schools.*

3.2 IDENTIFICATION DICTIONARIES AND ENCYCLOPAEDIA

17 Bean W J (1970) *Trees and Shrubs Hardy in the British Isles.* Rev. Edn. Murray.

18 Borg J (1959) *Cacti. A Gardeners Handbook for their Identification and Cultivation.* 3rd Edn. Blandford.

19 Clapham A R Tutin T G and Warburg E F (1962) *Flora of the British Isles.* 2nd Edn. Cambridge University Press.

20 Fitter R S R (1971) *Finding Wild Flowers.* Collins.

21 Hellyer A G L (1960) *Encyclopaedia of Garden Terms and Words.* Collingridge.

22 Hellyer A G L (1960) *Encyclopaedia of Plant Portraits.* Collingridge.

23 Jane F W Rev. (1963) *Conifers.* Special Leaflet No. 7. School Natural Science Society.

24 Keble Martin W (1965) *The Concise British Flora in Colour.* Ebury Press and Michael Joseph.

25 Makins F K (1957) *Herbaceous Garden Flora.* Dent.

26 McClintock D and Fitter R S R (1956) *The Pocket Guide to Wild Flowers.* Collins.

27 Perry F (1957) *Guide to Border Plants.* Collins.
28 Phillips G A R (1954) *The Book of Garden Flowers.* Warne.
29 Usher G (1966) *A Dictionary of Botany.* Constable.
30 Watson E V (1968) *British Mosses and Liverworts.* 2nd Edn. Cambridge University Press.
31 Welch H J (1966) *Dwarf Conifers.* Faber.
32 Willis J C (1966) *A Dictionary of the Flowering Plants and Ferns.* 7th Edn. Cambridge University Press.

3.3 SAFETY

33 Department of Education and Science (1967) *Safety at School.* Education Pamphlet No. 53. HMSO.
34 Department of Education and Science (1976) Safety Series No. 2. *Safety in Science Laboratories.* Revised Edition. HMSO.
35 Forsyth A A (1968) *British Poisonous Plants.* Ministry of Agriculture Fisheries and Food. Bulletin No. 161 2nd Edn. HMSO.
36 North P M (1967) *Poisonous Plants and Fungi in Colour.* Blandford Press.
37 Pharmaceutical Society of Great Britain (1966) *Common Poisonous Plants and Fungi.* Set of 35mm Colour Transparency Slides.
38 Royal Society for the Protection of Accidents (1964) *These Fruits are Dangerous* and (1966) *Poisonous Fungi.* Coloured posters.
39 Yates R J (1973) A Toxic Seed. *School Science Review* 54 (188) 511.

3.4 FACILITIES; CONTROL OF PESTS AND DISEASES

40 Agricultural Research Council Weed Research Organisation (1972) *Chemical Weed Control in Your Garden.*
41 Bickford E D and Dunn S (1972) *Lighting for Plant Growth.* Kent State University Press, Kent, Ohio.
42 Biological Science Curriculum Study. Barthelemy R E Dawson J R and Addison E Lee (1964) *Innovations in Equipment and Techniques for the Biology Teaching Laboratory.* D C Heath/Harrap.
43 Bingham C D (1968) The Culture and Use of Plants in School. *Journal of Biological Education* 2 (4) 353–64.
44 Canham H E (1964) *Electricity in Horticulture.* Macdonald.
45 Electricity Council (1971) *Electric Growing.*
46 Fox-Wilson G Revised Becker P (1960) *Horticultural Pests. Detection and Control.* Crosby Lockwood.
47 Gram E Bovien P and Stapel C (Eds) (1969) *Recognition of Diseases and Pests of Farm Crops.* 2nd Edn. Blandford.
48 Gray A W (1966) *Gardening with Electricity.* Royal Horticultural Society. Reprinted from the *Journal of the Royal Horticultural Society* (1958) 83 (4) with revisions.
49 Hellyer A G (1966) *Garden Pests and Diseases A Guide to Recognition and Control.* Collingridge.
50 Lawrence W J C and Newell J (1962) *Seed and Potting Composts.* 5th Edn. Allen & Unwin.
51 McLean R C and Ivimey Cook W R (1952) *Plant Science Formulae.* 2nd Edn. Macmillan.
52 Ministry of Agriculture Fisheries and Food (1964) *Capillary Watering of Plants in Containers.* HMSO.
53 Ministry of Agriculture Fisheries and Food (1964) *Commercial Glasshouses.* Bulletin 115 4th Edn. HMSO.
54 Ministry of Agriculture Fisheries and Food (1962) *Irrigation.* Bulletin 138 3rd Edn. HMSO.
55 Nuffield Foundation Science Teaching Project Advanced Science: Biological Science. Fry P J (Ed) (1971) *Laboratory Book.* Penguin.
56 Purvis M J, Collier D C and Walls D (1966) *Laboratory Techniques in Botany.* 2nd Edn. Butterworths.

3.5 MAINTENANCE, CULTURAL DIRECTIONS

57 Ashberry A (1964) *Bottle Gardens and Fern Cases.* Hodder & Stoughton.
58 Baker K (1972) *U.C. System of Producing Healthy Container Grown Plants.* University of California, Los Angeles.
59 Behringer M P (1973) *Techniques and Materials in Biology.* McGraw-Hill.
60 Clapham S (1972) *Primulas.* David & Charles.
61 Clifford D (1970) *Pelargoniums including the popular 'Geranium'.* 2nd Edn. Blandford.
62 De Wit H C D (1964) *Aquarium Plants.* Blandford.
63 Epstein E (1972) *Mineral Nutrition of Plants;*

Principles and Perspectives. John Wiley & Sons, New York.
64 Foster E G (1964) *The Gardener's Fern Book.* Van Nostrand.
65 Fraser H (1966) *The Gardener's Guide to Pruning.* Collingridge.
66 Gardiner G F (1957) *Pictorial Plant Propagation.* Pearson.
67 Garner R J (1967) *The Grafters Handbook.* 3rd Edn. Rev. Faber & Faber.
68 Hartmann H T and Kester D E (1968) *Plant Propagation Principles and Practices.* 2nd Edn. Prentice Hall.
69 Hay R (1965) *Annuals.* Bodley Head.
70 Hellyer A G L (1972) *The Amateur Gardener.* 3rd Edn. Rev. Collingridge.
71 Hellyer A G L (1960) *Sanders' Encyclopaedia of Gardening.* Collingridge.
72 Higgins V (1971) *Cactus Growing for Beginners.* New and Revised Edn. Blandford.
73 Jones M H (1962) *House Plants.* Penguin.
74 Lees L (1970) Ecological Gardens. *Biology and Human Affairs* 36 (1) 35–40.
75 Lewis C C (1965) *The Greenhouse.* Pergamon.
76 Macself A J (1965) *The Amateurs Greenhouse.* Collingridge.
77 Ministry of Agriculture Fisheries and Food (1970) *Commercial Production of Pot Plants.* Bulletin 112. 4th Edn. HMSO.
78 Moore H E (1957) *African Violets, Gloxinias and their Relatives.* Macmillan.
79 Park B (1962) *Collins Guide to Roses.* Collins.
80 Procktor N J (1968) *Simple Propagation.* Collingridge.
81 Robinson G W (1959) *The Cool Greenhouse.* Penguin.
82 Rochford T and Gorer R (1969) *The Rochford Book of House Plants.* Faber.
83 Royal Horticultural Society (1956) *Dictionary of Gardening.* 2nd Edn. and Supplement (1969). Oxford University Press.
84 Royal Horticultural Society (1965) *The Fruit Garden Displayed.* Oxford University Press.
85 Royal Horticultural Society (1961) *The Vegetable Garden Displayed.* Oxford University Press
86 Swindells P (1971) *Ferns for Garden and Greenhouse.* Dent.
87 Synge P M (1971) *Collins Guide to Bulbs.* 2nd Edn. Collins.
88 Walker L M (1971) *Bonsai.* John Gifford.
89 Walls I (1970) *Greenhouse Gardening.* Ward Lock.

3.6 USE

90 American Phytopathological Society, Sourcebook Committee; Kelman A, Chairman (1967) *Sourcebook of Laboratory Exercises in Plant Pathology.* W H Freeman.
91 Andrews J N and Hornsey D J (1967) Experiments with Radioisotopes in School Biology. *School Science Review* 48 (166) 748–58.
92 Baron W M M (1967) *Organisation in Plants.* 2nd Edn. Edward Arnold.
93 Boothroyd S W and Kelman A (1966) Laboratory Experiments in Plant Pathology. *American Biology Teacher* 28 (6) 478–502.
94 Branson J M (1967) *School Gardens and Nature Study.* In Publication No. 28 School Natural Science Society.
95 Clarke R A, Booth P R, Grigsby P E, Haddow J F and Irvine J S (1968) *Biology by Inquiry* Book 1 Students Text (also Teachers Guide). Heinemann.
96 Clevenger S (1964) *Flower Pigments.* Scientific American Offprint No. 186.
97 Davis B D and Postlethwait S N (1966) Classroom experiments using fern gametophytes. *American Biology Teacher* 28 97–102.
98 Devon Trust for Nature Conservation (1972) *School Projects in Natural History.* Heinemann.
99 Educational Use of Living Organisms Project (Schools Council) (1976) Comber L C *Organisms for Genetics.* Hodder and Stoughton Educational.
100 Finch I E (1964) *Trees—their Form and Branching.* Special Leaflet No. 21 School Natural Science Society.
101 Freeland P W (1972) G A Enhanced a-Amylase Synthesis in halved grains of Barley (*Hordeum vulgare*): a simple Laboratory Demonstration. *Journal of Biological Education* 6 (6) 369–75.
102 Hancock J E (1960) *The School Garden* Books 1–4 Macmillan.
103 Hannay J W (1967) Light and seed germination—an experimental approach to photobiology. *Journal of Biological Education* 1 (1) 65–73.

104 Heinowski Sister M R (1967) Simple Apparatus for Uptake of $C^{14}O_2$ in Aquatic and/or Terrestrial Plants and Autoradiography. *American Biology Teacher* 29 (1) 30–3.

105 Keefe A M (1965) Plants for a Window Garden *American Biology Teacher* 27 (2) 118–123.

106 Mylechreest M (969) *Cuscuta Campestris. School Science Review* 50 (172) 570.

107 Mylechreest M (1971) The Secondary School Garden. *School Science Review* 53 (182) 77–84.

108 Nuffield Foundation Science Teaching Project Advanced Science: Biological Science. Stoneman C F (Ed) (1970) *Maintenance of the Organism: A Laboratory Guide.* Penguin.

109 Nuffield Foundation Science Teaching Project Advanced Science: Biological Science. Gray J H (Ed) (1970) *Organisms and Populations: A Laboratory Guide.* Penguin.

110 Nuffield Foundation Science Teaching Project Advanced Science: Biological Science (1970) *Teachers Guide Maintenance of the Organism. Organisms and Populations.* Penguin.

111 Nuffield Foundation Science Teaching Project Advanced Science: Biological Science. Sands M K (Ed) (1970) *The Developing Organism: A Laboratory Guide.* Penguin.

112 Nuffield Foundation Science Teaching Project Advanced Science: Biological Science. Barker J A (Ed) (1970) *Control and Co-ordination in Organisms. A Laboratory Guide.* Penguin.

113 Nuffield Foundation Science Teaching Project Advanced Science: Biological Science (1970) *Teachers Guide 2 The Developing Organism: Control and Co-ordination in organisms.* Penguin.

114 Nuffield Foundation Science Teaching Project Biology (1966) *Text 1: Introducing Living Things* (also *Teachers Guide 1*). Longmans/ Penguin.

115 Nuffield Foundation Science Teaching Project Biology (1966) *Text 2: Life and Living processes* (also *Teachers Guide 2*). Longmans/ Penguin.

116 Nuffield Foundation Science Teaching Project Biology (1966) *Text 3: The Maintenance of Life* (also *Teachers Guide 1*). Longmans/Penguin.

117 Nuffield Foundation Science Teaching Project Biology (1969) *Text 4: Living Things in Action* (also *Teachers Guide 4*). Longmans/ Penguin.

118 Nuffield Foundation Science Teaching Project Junior Science (1967) *Animals and Plants.* Collins.

119 Nuffield Foundation Science Teaching Project Secondary Science (1971) *Theme 1: Interdependence of Living Things.* Longmans.

120 Nuffield Foundation Science Teaching Project Secondary Science (1971) *Theme 2: Continuity to Life.* Longmans.

121 Nuffield Foundation Science Teaching Project Secondary Science (1971) *Theme 6: Movement.* Longmans.

122 Nuffield Foundation Science Teaching Project Secondary Science (1971) *Teachers Guide.* Longmans.

123 Paice P A M (1968) Simple Radioisotope Experiments in School Biology. *School Science Review* 49 (170) 62–78.

124 Prime C T (1971) *Experiments for Young Botanists.* Bell.

125 Rhodes L W (1968) The Duckweeds—Their Use in the High School Laboratory. *American Biology Teacher* 30 (7) 548–551.

126 Science 5/13 (1973) *Trees.* Macdonald Educational.

127 Scottish Schools Science Equipment Research Centre (1972) Transpiration in Plants. *Bulletin* 58 2–7.

128 Shove R F (1972) *The Uses of Trees Cultivated in Pots for the Study of Plants in School.* Publication No. 2. 3rd Edn. School Natural Science Society.

129 Syrocki J (1971) Grafting in Herbaceous Stems. *American Biology Teacher* 33 (9) 532–4.

130 Thomason B (1972) Bryophyte Gemmae for Class Practical Work. *School Science Review* 53 (184) 560–1.

131 Vinnicombe E J (1972) *How to Make Prints and Plaster Casts of Plant Material.* Publication No. 17. School Natural Science Society.

132 Wilson R W (1970) *Making a Collection.* Publication No. 39. School Natural Science Society.